Synthesis Lectures on Power Electronics

Series Editor

Jerry Hudgins, Lincoln, USA

This series publishes short books on topics related to power electronics, ancillary components, packaging and integration, electric machines and their drive systems, as well as related subjects such as EMI and power quality. Each Lecture develops a particular topic with the requisite introductory material and progresses to more advanced subject matter such that a comprehensive body of knowledge is encompassed. Simulation and modeling techniques and examples are included where applicable.

Shivkumar V. Iyer · Mohammad Nair Aalam

Switching Strategies for Power Electronic Converters

Examples with Python Simulations

 Springer

Shivkumar V. Iyer
Hamburg, Germany

Mohammad Nair Aalam
Virginia Tech
Blacksburg, VA, USA

ISSN 1931-9525   ISSN 1931-9533 (electronic)
Synthesis Lectures on Power Electronics
ISBN 978-3-031-41404-6   ISBN 978-3-031-41405-3 (eBook)
https://doi.org/10.1007/978-3-031-41405-3

This Springer imprint is published by the registered company Springer Nature Switzerland AG
The registered company address is: Gewerbestrasse 11, 6330 Cham, Switzerland

Paper in this product is recyclable.

In the memory of my late mother
Shivkumar V. Iyer

For my parents
Mohammad Nair Aalam

Preface

Power electronic converters have become ubiquitous in modern times with the advent of smart grids, renewable energy sources and electric vehicles. This in turn has led to a rapid expansion of the power converter market with numerous companies offering a wide range of power converters for different applications. In order to sustain this growing market, the industry needs fresh young engineers who are well trained in the operation and control of power converters. Though universities have stepped up to make their courses available to the general public and not just their own students, there is still a need for detailed and stand-alone courses that serve as educational material for the next generation of power engineers.

This book is a part of a larger project that aims to provide accessible and affordable education to young engineers who wish to specialize in power electronics. This book deals with how switching strategies can be formulated for power electronic converters that are popularly used in the power industry. The approach used in the book uses logical reasoning and the basic principles of engineering to examine how power converters can be provided with gating signals for various applications. The book begins with the absolute basics, describing the philosophy behind Pulse Width Modulation and demonstrating frequency analysis tools, and progresses on to fairly complex topics such as how Space Vector Pulse Width Modulation can be implemented for a multi-level converter.

This book extensively uses simulation as a learning tool, with every switching strategy accompanied by a simulation. Moreover, by using the free and open source circuit simulator Python Power Electronics, the simulations in the book are universally accessible to a reader in any part of the world. The book also contains suggestions on how interested readers can extend the simulations provided for their own projects or according to their own interests. All the programs and simulations used in the book can be found in an online repository and can be freely downloaded by interested readers in any part of the world. Interested readers can also find numerous free resources for learning power electronics by visiting the homepage of Python Power Electronics.

Hamburg, Germany   Shivkumar V. Iyer
Blacksburg, USA   Mohammad Nair Aalam

Acknowledgments

I would like to thank my father and my sister for their support and encouragement in writing books. I would like to thank Prof. Bin Wu for being a supportive mentor as I embarked on becoming an author of technical books. And my greatest thanks is reserved for Rakhat for her constant support, encouragement and patience as I spent countless hours on this book project.

Shivkumar V. Iyer

Alhamdullilah. I would like to extend my thanks to my parents and my sister for their unwavering support throughout my innumerous endeavors. I am grateful to Shiv for his teachings and for providing me the opportunity to collaborate on this book project.

Mohammad Nair Aalam

Contents

Introduction

1

1.1   Challenges in Power Electronics Education

In the past decade, power electronics has steadily emerged as one of the most popular specializations in electrical engineering displacing others such as power systems, high voltage systems, and switchgear and protection. This has given the power electronics industry some hope as among those students who choose to continue in core electrical, the majority choose to become power electronics engineers. However, this statistic pales in front of the statistic that shows the exodus of engineers from the hard engineering domains such as electrical, mechanical, chemical and civil towards Information Technology (IT), software and many other soft engineering domains. Taking into consideration that the software domain has not at all come close to saturation, but rather has been continuously growing in its demands for young engineers as global business goes completely digital, the power industry in general has been growing increasingly nervous.

Artificial Intelligence (AI) and Machine Learning (ML) are now by far the most popular domains for any young person inclined towards tech. With major advances announced by global multinationals several times a month, it is very clear that the excitement that these domains offer to those who are about to graduate is unmatched by any other domain of engineering. The vast majority in core engineering no longer even think of competing with the IT industry in trying to attract talent as they feel it is a war that has already been lost. Not very long ago, I used to see posts on professional networking sites encouraging youngsters to choose the core engineering domains rather than merely think of IT. In recent times, I see posts related to the lack of engineers in the core domains that have turned the matter into a joke making it clear that core engineering is being seen as a lost cause.

Even if one accepts the fact that the power electronics industry can never truly match the IT industry, and Electric Vehicles (EVs) can never produce the same level of excitement that AI can, it is also important to admit that in terms of educational opportunities, core engineering has done a poor job as compared to the IT industry. If core engineering has

© The Author(s), under exclusive license to Springer Nature Switzerland AG 2024   1
S. V. Iyer and M. N. Aalam, *Switching Strategies for Power Electronic Converters*,
Synthesis Lectures on Power Electronics, https://doi.org/10.1007/978-3-031-41405-3_1

any hopes of surviving in the next decades to come, it must learn a few lessons from the IT industry that are not very difficult to implement. Most of these lessons are in the domain of education, as unless young people have convenient access to learning materials, very few will choose to stay where they feel they may stagnate. In the rest of this section, I will focus on a few aspects of the IT industry that have enabled it to surge ahead in popularity in the past couple of decades.

Many decades ago it was customary for an engineer to find work in his or her domain of study. Those who worked in another domain from the one they graduated in were usually the odd ones out, and in most cases had special reasons for changing their domain. All this began to change in the 1990s when the software industry began to experience a boom. The software industry welcomed all engineers with open arms. Though there were a few niche applications in the software industry for which engineers were still expected to have a Computer Science degree, for most other applications, this requirement was relaxed. To enable engineers from other domains to find their feet in the IT industry, software companies went to great lengths in terms of training programmes. This need to train non-software engineers in programming and other related skills such as server administration, operating systems and many others led to the availability of online learning material that was universally accessible.

Fast forward to these modern times, anyone who wishes to learn any topic related to IT can find numerous resources, ranging from completely free to those that are as expensive as university degrees. Though such online learning resources are gradually becoming available in core engineering, they have not been accepted to the same extent by engineering companies as they have been accepted by software companies. Most engineering companies still screen their candidates using the traditional processes. Therefore, if one steps back and takes another look at the problem of scarcity of engineering in core domains, one must ask the question— are core engineering companies doing enough to provide learning resources the way the software industry did a few decades back? The answer unfortunately is no—these companies are still hoping that universities alone will fulfil the role of educating young talent.

There are many reasons why core engineering companies are reluctant to change their view of education, some are quite obvious while a few are much more subtle. The most obvious reasons are the costs involved in core engineering education compared to the costs of online learning in IT. In IT, a few technologies need little more than a computer and an internet connection for a newcomer to become well-versed with. A few other technologies may need additional investment in terms of purchasing servers or even hardware such as Graphical Processing Units (GPUs) for those seeking to specialize in AI. In comparison, in-depth learning in core engineering quite often needs significant investment in terms of infrastructure, laboratories, machines and equipment. Though in these modern times, one can learn a good deal through simulations, simulation-based learning is still considered by many to be substandard compared to learning based on theory and experiments. This unwillingness to embrace new learning models is unfortunately the reason why online learning in core engineering is still far behind equivalent online courses in IT.

Though one can accept that to become a practising engineer, it is necessary to be well-versed with theory and also possess good practical knowledge, one cannot merely focus on the end without giving any thought to the means that are necessary to achieve the end. As we embark on the journey of becoming engineers, a number of different experiences serve to increase our interest and also give us the momentum to persevere in this journey. Though for the industry, their final interest is well-trained engineers who graduate, for the engineers who are undergoing training, a variety in the learning experiences as well as a flavour of technology from different perspectives will make a big difference in piquing the interest of students. Though traditional education offers depth and rigour, it unfortunately is unable to offer this variety, and this in turn leads to student disinterest.

In the IT industry, online learning is at times convenient while at other times necessary. By providing low-cost and sometimes free learning resources, young people were able to gain exposure to programming and software. Even if one argues that such skills may not be of direct use in industry, the level of comfort that it produces in those who are exposed to these online resources results in a greater interest in joining the IT industry. Therefore, the convenience of online learning resources has merely formed a bridge into the industry from where rigorous skills can be learned through various resources. At other times, due to the rapid advancement of technology in the industry, one has no option but to seek out informal learning resources, as conventional learning resources are not yet available. Therefore, the necessity of indulging in online learning makes these resources more acceptable on the whole.

Therefore, to sum up the challenge that is faced by the core engineering industry which also includes power electronics, is that education needs to be more flexible, more diverse and should appeal to young people. The last factor is by far the most important, as it is extremely important to reach out to young students who are in their junior undergraduate years with interesting courses and projects—to "catch them young". For this purpose, education must drastically change, as the current mode of teaching in core engineering is far too dull to captivate the mind of a young person. Luckily, power electronics is a domain that borrows from several other domains—signal processing, control systems, embedded systems, communications and many others. Therefore, to be able to structure interesting course material that combines power electronics with many of these other interesting domains should not be too difficult.

If one were to ask established power electronics engineers what made them choose power electronics as a specialization, the answer will rarely be stable employment or good income. Typical answers will be that one project they stayed up all night working on, or a professor whose lectures simply blew them away and many more. Therefore, for the power electronics industry to fulfil its need for young engineers, all it has to do is ignite the curiosity of engineers when they are young. To merely hope that engineers will miraculously find the interest to work in power electronics is to be living in a fool's paradise. The next few sections will describe how this book in a part of a larger series of educating young power engineers and awakening the "tinkerer" in them.

1.2   Meeting the Requirements of Industry

In the previous section, we stressed on the need for revamping education in power electronics in order to continue attracting young talent. Since most engineers in core engineering domains attribute the dearth of qualified engineers to the exodus of talent to the IT industry, the section attempted to examine how the IT industry managed to drastically change educational opportunities for young engineers. In this section, let us continue this discussion by examining at an abstract level what skills are required in the power electronics industry and how young engineers can acquire them. The initial discussion will be on the abstract level as when one seeks to make any changes, it is important to be able to differentiate between changes that will result in a major impact versus those that will merely be disruptive but will not result in any significant gain.

Since the objective of restructuring power electronics education is to encourage them to join the power electronics industry, new educational resources will have to be tailored specifically according to the needs of industry. Therefore, the first question one needs to ask is—what does industry want? This question is surprisingly difficult to answer. If left to industry to answer this question, the result will be almost impossible to achieve, as industry is looking for incredibly bright motivated engineers who can complete projects with absolutely no supervision, and also have the soft skills to document and present the final work. Therefore, rather than pose this question arbitrarily to someone in the power electronics industry, it would be more prudent to perform an analysis of the different type of job positions that exist in the industry, and how engineers can be best trained for these positions.

Let us examine a fundamental unit of a power electronics company as one that produces a set of products for a particular application. These units could either be companies themselves, or could be a division of a large company. Examples could be a company that produces Uninterruptible Power Supplies (UPS) as backup power supplies for either homes, offices or even for industries, or a company that produces Variable Speed Drives (VSD) to control motors for industrial applications. Usually, such companies will offer a range of products for different applications, and over time will add newer products to their portfolio while also rendering some products obsolete. If one examines the manpower requirement of such companies, they require engineers to install their products, inspect their products, service their products, repair and test their products, and also design newer products. Installation, inspection and servicing products usually occurs at client locations and is performed by specialized and trained teams. These teams will not only have the necessary knowledge about the company's products, but will also be trained in Environment, Health and Safety (EHS) to ensure that installation follows safety standards. Such teams are intensively trained by the company, and therefore, one does not have to think of novel education strategies for these engineers.

The remaining two functionalities within the company are the need to repair and test products, and to design newer products. These tasks are usually performed in the company premises, and the skills needed by the engineers can vary to a great deal. In terms of repair and testing, engineers need to examine appliances that have been found defective and returned by the client. Though this may seem routine and mundane, at times, this can be fairly challenging especially in the case of an appliance such as a UPS or a VSD. An appliance that comprises at least one power electronic converter can fail due to many reasons—failure of any one power device or component, failure of control circuitry or failure of auxiliary circuitry such as protection or communication systems. To be able to identify the cause of the failure and rectify not just the failed component, but also all other components affected by the failure, requires a good deal of knowledge not only about components, power electronics and converters, but also about control systems and basic electronics.

The final task of designing newer products is quite often better known as Research and Development (R&D). Designing newer products requires the drawing up of product specifications with respect to existing company products, competitor products as well as unfulfilled requirements in the larger industry. The new product is designed and analysed using specialized tools, following which prototypes are fabricated that are tested. The entire life cycle of new product creation can take several months to several years depending on the complexity of the appliance. The engineers involved in this process quite often have advanced engineering degrees, and advanced knowledge of power electronics, control systems and many other domains that results from previous involvement in research projects. It is also important to note that not all research projects result in newer products, as quite often, the advantages offered by the new product may not be sufficient to replace an existing product. In such cases, research projects advance the Intellectual Property (IP) held by the company which can be licensed out by other companies.

One could think of many other job functions that exist in companies. However, as with installation and service, many functions are very specific to the company or industry, and need specialized training for which industries invest heavily. In most cases, the general engineering staff that are hired by companies are usually hired for testing and repair, or in R&D. Therefore, if one targets these two functions found in most companies, and examines how engineers can be trained for these roles, we now have a far more concise job description. Most of the discussion that will follow is the result of analysing commonly found job openings in different companies, and though companies will have differing requirements based on the technology that they use, it is easy to see some common threads among different companies.

One family of job descriptions that are found are those of Test Engineer, Validation Engineer or even Test Validation Engineer. These jobs require engineers to either manually test appliances especially those that have been returned as defective by the customer, or to setup automatic testing platforms for appliances that are being produced by the company with the objective of performing Quality Assurance (QA) testing. While manually testing appliances, an engineer will need basic hardware skills to extract or replace components. In addition, the extracted components will need to be tested according to specifications, and if they are control circuitry or microcontroller based systems, may need to be verified or reprogrammed. In order to setup an automatic test environment, an engineer will need knowledge of automation software, embedded systems and programming languages for setting up automatic scripts. The purpose of these automatic tests is to generate a set of known inputs and verify the behaviour of the appliance against expected outputs. Many of these positions will require the engineer to be well-versed with a particular automation software, besides also possessing good programming skills. Though programming skills can be easy to acquire through online courses, experience with specific automation software might be a bit difficult especially when the software is proprietary, in which case, the training would need to take place in the company.

Another class of job descriptions quite often found are those of Hardware Engineer or Firmware Engineer. In some smaller companies, these are also combined as Design Engineer. These jobs are surprisingly abundant, and engineers in these position perform a variety of tasks in fabricating prototypes usually for newer products. These tasks can include designing Printed Circuit Boards (PCBs), designing basic electronic circuits, writing control code for microcontrollers and other hardware-based controllers, and also manual and automatic testing of hardware prototypes. These engineers can also serve as junior engineers in R&D teams, who primarily implement and test the designs created by senior engineers using advanced tools and techniques. Such a division of labour usually increases the efficiency of the R&D department, as senior researchers can step away from the hardware implementation and focus on advanced design challenges, while junior engineers can focus on hardware implementation without being faced with the challenges of design which can be quite complex.

There exist many other job functions within companies, but these two are usually the junior-most, and are best suited as entry-level jobs for engineering graduates looking for their first positions in industry. With that said, how would one prepare engineers for these jobs in the simplest and most affordable manner while still instilling an interest in engineering? One of the simplest skills to learn in these modern times is programming. There are numerous online courses for programming in various contexts—general programming courses that solve mathematical problems, programming for scientific applications, programming for electronics hobbyists and many more. These programming skills can not only be useful, but can be immensely enjoyable, as most of these courses can be combined with low-cost projects that can be completed by young people from the comfort of their homes. Besides enhancing skills, these courses perform wonders at igniting creativity in young people.

The next level of skills that can be acquired by young engineers through online courses is on basic electronics. A number of low-cost and low-power projects can be completed for a number of minor applications such as light dimmers, miniature motor-speed regulators and other home automation applications. Engineers can also learn how to use electronics circuit simulation software, as many of the online courses use free and open-source software which are free from licensing requirements. Besides circuit simulation, engineers can learn how to design PCBs for their electronic circuits, such that the final prototype need not be a tangle of wires on a breadboard, but rather can have the appearance of a professional product. Nowadays, PCBs can be designed using free and open-source software, and the final designs can be fabricated for a relatively low cost by many fabrication workshops that specifically cater to electronics hobbyists. The absolute thrill of designing an electronic circuit, transferring the design to a PCB and testing out the final circuit, is something most electronics hobbyists will state as the reason for their continuing in their tinkering journeys.

Finally, we can arrive at the next level of skills—understanding power electronic converters. To become well-versed with power electronics, though it is now possible to find online learning resources, most of them fail to ignite the curiosity of young minds, as these are merely university courses made online. It is at this juncture that the power electronics industry faces an obstacle—to be able to find online learning resources that are as engaging as programming courses or basic electronics courses. Unfortunately, a part of the problem rests with finding people with adequate skills and knowledge to be able to take the time to create such courses. Most who possess these skills are either in very senior positions in industry, or are senior professors in universities, and being pressed for time, are unable to create sufficiently engaging courses for young engineers.

With this overview of the requirements of industry, it is quite easy to state that the challenge for the power electronics industry is the ability to create engaging educational content in advanced topics as the technology progresses. It is important to note that not only are such forms of educational content necessary to train new engineers entering the industry, but also to enable existing engineers who may be in junior positions to upskill and get promoted to senior positions. Such a fluid career path within the power electronics industry will make the industry attractive to young engineers. The next section will describe how this book is part of a larger project in educating young engineers in basic as well as advanced topics of power electronics.

1.3   Approach Used in the Book

The previous section highlighted the challenges faced by the power electronics industry with respect to educating young engineers. The section described in brief some of the popular job positions available as entry-level jobs to junior engineers who are entering the power industry. Engineers can prepare themselves for these jobs by learning programming and basic electronics through numerous online resources. However, learning resources for power electronics are still scarce and are merely online versions of university courses. This section will describe how this book and the larger project that this book belongs to, will attempt to provide learning resources in power electronics that can be as flexible and imaginative as the resources available in the IT domain.

As undergraduate students in electrical engineering begin with their power electronics courses, they are struck by one stark contrast with respect to the analog electronics courses taken before—the scarcity of equations that define the precise operation of the system. In most analog electronics, one can calculate to a fair degree of precision the exact operating condition of the circuit. This is primarily due to the fact that analog electronics primarily operate in the active region due to which transistors and other devices can be represented by current gains. Power electronic circuits on the other hand consist of power devices that work either in the saturation region or the cut-off region, behaving very similar to switches that can conduct a current in only one direction. Therefore, the analysis of power electronic circuits is usually performed by examining different conduction paths for different time segments during which different power devices conduct. Though it enables us to understand the working of the circuit using network equations and waveforms, the behaviour of any power electronic circuit seems random and heuristic to any newcomer to the domain of power electronics.

Unfortunately, most textbooks do not address this confusing aspect of power electronics. The confusion only increases as one progresses to more complex topologies, where switching strategies used are presented without an explanation of why they are necessary or sufficient. For a student who through high-school and the early undergraduate years has been taught to swear by mathematics and analysis, power electronics seems arbitrary and without any fundamental mathematical basis. Quite often, these students move onto domains such as communications or signal processing which are mathematically intense. As stated in the previous section, this inability to retain bright young engineers in power electronics eventually leads to a scarcity of skilled talent in the power electronics industry. In the past decade, power electronics courses have been taught with simulations as an accompanying tool, and students simulate converters besides also completing projects as a part of their course requirements. However, despite the fact that simulations add to a certain degree of interactivity in the learning process, where students can try things out and examine the results, for these simulations to be truly effective, the presentation of the learning material needs to drastically change.

This book is a part of a series of books on teaching power electronics through simulations. This book differs from other books in power electronics. To begin with, most books on power electronics are accompanying texts to university courses, and cover a vast range of topics which would form the syllabus of any power electronics course at the university level. This book on the other hand, is presented as a standalone material with which a student can learn only a select few topics in power electronics. Rather than attempt to cover a wide range of topics, the book focuses only on a narrow collection of topics, but instead, presents a extremely deep examination of these topics. Since the entire focus on this modified learning approach is to prepare students for industry, the collection of topics is directly relevant to applications in industry. In other power electronics textbooks, simulations are used as accompanying material. However, in this book, simulations are an integral part to explaining every concept. Every circuit and every switching strategy is accompanied by a simulation, and also with an in-depth analysis of the simulation procedure and results. This is to ensure that students develop a rigorous understanding of the process of simulation, and are able to simulate their own circuits.

Students who are reading this book can use any circuit simulator that they may have access to. The circuit simulator being used must allow the user to create a circuit schematic with controllable power devices resembling Metal Oxide Semiconductor Field Effect Transistors (MOSFETs) or Insulated Gate Bipolar Transistor (IGBT) and uncontrollable power devices resembling diodes. Furthermore, the circuit simulator must allow the user to create a custom control file to generate gating signals to the controllable power devices to verify switching strategies. The circuit simulator must allow the reader to plot the currents in different parts of the circuit, and the voltages across different branches of the circuit. Furthermore, for in-depth analysis of the switching strategies, the circuit simulator must also allow the user to combine multiple quantities in a single plot. A vast number of proprietary and free circuit simulators will be able to perform this task.

This book will use the free and open-source circuit simulator Python Power Electronics. Python Power Electronics has been written completely using Python which is a free and open-source high level programming language. Python Power Electronics has been written especially for power engineers who wish to simulate power electronic circuits while implementing control digitally. The homepage of Python Power Electronics can be found at: https://www.pythonpowerelectronics.com/.

The circuit simulator can be downloaded from the link: https://www.pythonpower electronics.com/?page=softwaredownloads.

The circuit simulator can be used through a web browser by launching a server on the computer. To learn how to install the circuit simulator and on how to use it to simulate power electronic circuits with digital control, the reader can view any of the video series on the following link: https://www.pythonpowerelectronics.com/?page=videos.

The reader need not use Python Power Electronics to perform the simulations in this course. Most of the code samples and simulations can be achieved using any other circuit simulator. The purpose behind using a free and open-source circuit simulator in this book is to promote the use of free and open-source software in electrical engineering. Free and open source software not only increases the accessibility of software but also builds a community of users that enables the sharing of information and knowledge. The reader is encouraged to follow the project on any number of different social media where regular updates are made. For example, a number of video lecture series can be found on YouTube on the video link above, where different topics in electrical engineering are explored using simulations.

All the simulations in this book have been hosted on GitHub to be freely accessible to everyone. The link for the repository that contains the simulations is: https://github.com/opensourceelectrical/switching-strategies-for-power-electronics.

The repository contains folders for every chapter. Within each folder are the simulations for that chapter. Each simulation has been arranged in a separate folder and there can also be nested folders in the case of separate cases being simulated. Every simulation will have a README file. The reader must first read the README file which describes what the simulation is about, what are the circuit schematic files used and what are the control files used. In the later chapters, as we describe each simulation, the exact files will be referenced and described in detail.

The reader is strongly encouraged to perform the simulations while reading this book and use the simulations provided in the GitHub repository as a reference. The simulations have been performed in such a manner that the switching strategies can be transferred to a microcontroller in a hardware implementation with the least amount of modifications. While describing the simulations, potential modifications to the simulations will also be suggested, and the reader is strongly encouraged to try these out as well. Simulations are a safe platform to try out circuits without the fear of circuits blowing up. This gives several opportunities for the learner to look "under the hood" and also poke around with simulation models.

1.4   Outline of the Book

The past few sections described why this book was written. The sections described how the challenges faced by the power electronics industry can be solved in the most effective manner by revamping education, and how this book is a part of a project that attempts to bring a new flavour into power electronics education. The readers are encouraged to take a look at some of the resources available in the Python Power Electronics project homepage provided in the previous section. The readers should specifically go through the resources related to downloading, installing and using the Python Power Electronics circuit simulator. This section will provide an overview of the book and describe the significance of each chapter.

Chapter 2 titled "The Concept of Modulation" introduces Pulse Width Modulation (PWM) used in power electronics in comparison to Amplitude Modulation (AM) and Frequency Modulation (FM) used in communications and signal processing. Though the applications are vastly different, the underlying philosophy behind modulation is the mixing of signals to overcome challenges in dealing with the signal that is available. In communications, in order to be able to transmit an audio signal that could be potentially weak and randomly varying, it is mixed with a high frequency signal called as a carrier. In power electronics, in order to be able to produce a low frequency output voltage from a converter comprised of non-linear power devices, the low frequency signal resembling the output voltage is compared with a high frequency carrier signal to generate pulses of variable width. These pulses are fed to the power devices as gating signals, eventually resulting in the converter producing a high frequency switched voltage. The chapter introduces frequency analysis techniques such as Fourier series, Fourier transform and Discrete Fourier Transform (DFT) to examine the frequency spectrum of the modulated signals. The chapter presents a buck converter as a sample power electronic converter to describe how PWM results in an output that can be conveniently filtered to produce a steady output voltage for an appliance.

With the basics of modulation described in Chapter 2, Chapter 3 titled "Coordinated modulation strategies—converters with two controllable power devices" describes how these modulation strategies will need to be modified for power converters with two controllable power devices. The chapter uses the example of a modified buck-boost converter to describe how the buck and boost mode of operation will need separate coordinated switching of the power devices. The chapter then introduces the converter leg which is very popular in power electronics and is commercially available as a module supplied by many manufacturers. The chapter describes how the converter leg can be used to realize basic topologies such as the buck, boost and buck-boost topologies while also enabling these converters to be bidirectional. The chapter introduces the concept of forbidden modes of conduction in the context of converter legs as with multiple controllable devices connected in series, they cannot all conduct simultaneously as it will result in a short-circuit of the dc voltage source connected across the converter leg. The chapter demonstrates how the converter leg is a fundamental building block in power electronics with basic dc-dc converters and also a half-bridge dc-ac converter.

Chapter 4 titled "Full-bridge Converter" presents the full-bridge converter consisting of two converter legs with a dc voltage source connected across them. This full-bridge converter is one of the most basic functional converters that can be used for a number of applications both in dc systems as well as in ac systems. The chapter describes the operation of the converter by examining all possible modes of conduction. Due to the use of converter legs, the conduction of all devices in either or both legs remains a forbidden conduction mode. However, using simple permutations, one can compute the number of distinct conduction modes that are possible, and the output voltage produced for each conduction mode. With this analysis, the chapter introduces the concept of vector representation of the converter voltage output. The chapter describes how depending on the application, it is possible to synthesize a certain pattern for the required output voltage, which in turn can be mapped to transitions between the output voltage vector. The chapter describes various applications for the full-bridge converter—bidirectional dc-dc converter and bidirectional dc-ac converter. Though it is possible to formulate a switching strategy using output voltage vectors, the chapter describes how carrier-modulation signal comparison based PWM can be modified for different applications.

Chapter 5 titled "Three-Phase Converters" continues with the usage of the converter leg as a building block in three-phase ac systems. Using the same approach of eliminating forbidden conduction modes and using permutations to determine the total number of allowable conduction modes, the chapter lists the output voltages produced by the converter for each conduction mode. To describe the operation of the three-phase converter, a simple simulation with carrier-modulation signal comparison based PWM is presented. The chapter then describes how any three-phase quantity can be represented as a vector using a very popular transformation called the Clarke's transformation. Using Clarke's transformation, the output voltages corresponding to every allowable switching combination can be transformed into a vector, and all the voltage vectors that can be produced by the converter can be represented on a vector diagram. Using this visual depiction of the converter output voltages, the chapter will then present the Space Vector Pulse Width Modulation (SVPWM) switching strategy through which converter voltage vectors can be chosen such that the output voltage is closely regulated to a desired reference. The chapter describes in detail the various considerations that need to be taken into account while choosing the converter voltage vectors, and also in calculating the time intervals for which they need to be applied. The chapter presents simulation results with the three-phase converter switched by SVPWM.

Chapter 6 titled "Multi-level Converters" describes a modification of power converters that become necessary for high power industrial and power system applications. The chapter describes how the level of any converter can be defined, and with this definition, describes the difference between 2-level converters described in the previous chapters and higher level converters. The chapter describes an intuitive manner in which the converter leg of the previous chapters that produced only two levels can be modified to produce multiple levels. The chapter then establishes an algorithm for constructing a multi-level converter leg using the Neutral Point Clamped (NPC) approach. Due to the larger number of controllable devices in a single converter leg, the chapter describes the operation of the converter leg in a logical manner, while also establishing the rule for defining forbidden conduction modes. The chapter uses the approach of permutations to establish the allowable conduction modes for a 3-phase 3-level and for a 3-phase 4-level converter, and also to compute the output voltages for each conduction mode. Using Clarke's transformation from the previous chapter, these output voltages can be transformed into vectors and visually represented in vector diagrams. The chapter then describes a systematic approach to choosing the converter voltage vectors such that the output voltage is closely regulated to a desired reference. The chapter finally presents a simulation of a 3-phase 3-level converter.

Chapter 7 concludes the book. The chapter describes the significance of each chapter, and what the reader should extract from them. The chapter describes the philosophy of the book and the larger project that the book belongs to, and provides the reader with additional links for more information.

The Concept of Modulation

2

2.1 Introduction

The purpose of this chapter is to gently introduce the concept of modulation in the power electronics domain. The reason for the use of the word "gentle" is that in most power electronics text books and courses, modulation is directly introduced in the context of power electronic converters, without a background as to why modulation is recommended. For this purpose, we take help from basic signal processing and communications, where some very basic examples can be presented to describe the process of modulation. Though the applications are drastically different, there is a common obstacle that modulation in both domains is trying to accomplish—how can a low frequency signal be put through a "process", and extracted at the end with minimal distortions. In the case of communications, the process is transmission over long distances, while in power electronics, it is shaping an output voltage or current in a desired form for an application.

The chapter will avoid going deep into signal processing or communication theory as it is extremely vast and not applicable to power electronics. We will examine the most basic mathematical expressions that describe the process of modulation, and we will implement it using Python simulations. Stress will be placed on those concepts that are easily interpretable without detailed references to signal processing. The chapter will include a basic description of frequency analysis concepts such as Fourier Series and Fourier Transform. As before, the theory behind these frequency analysis techniques is extremely vast. The chapter will present these techniques in the most basic form useful for a power electronics engineer, and with references to the basic simulations that have already been presented.

The chapter will then describe how in-built methods in Numpy can be used to achieve frequency analysis using just a few commands. However, though these commands can be found in the documentation of Numpy, there is a great deal of confusion about how they should be used. In this chapter, the use of these methods is compared with the theory behind the frequency analysis techniques, so that the reader can alter these for any other application.

© The Author(s), under exclusive license to Springer Nature Switzerland AG 2024   13
S. V. Iyer and M. N. Aalam, *Switching Strategies for Power Electronic Converters*,
Synthesis Lectures on Power Electronics, https://doi.org/10.1007/978-3-031-41405-3_2

Moreover, the chapter will also describe how the frequency spectrum plots produced by these in-built methods can be interpreted with respect to theory. Several simulations will be presented for the reader to examine all the variations that can result in frequency spectra.

The chapter will gradually introduce modulation in the power electronics domain. Initially, this will be done in an abstract manner without any power converter. After a basic verification of the effectiveness of modulation, the buck converter will be presented for implementing our learnings of modulation. A basic description of the buck converter is provided so that the reader can understand the challenge involved in regulating the operation of the buck converter, and how it can be overcome through modulation. The simulation results will clearly show how modulation results in a smooth waveform. The results are additionally verified through a frequency analysis on the simulation data.

2.2   Applications in Communications

Before we begin to describe the process of modulation in power electronics, let us examine how modulation has been used in other domains, as some of those aspects of modulation are simpler to understand. As electrical engineers, most of us have enrolled for courses in the basics of signal processing in our junior undergraduate years. These courses typically cover the basics of communications, though details are reserved for specialized courses that are usually offered in the senior years. For the discussion that follows, we only need the very basics, as it will be quite evident that there are strong similarities between the modulation used in communications and the modulation used in power electronics, despite the fact that the applications are drastically different. Furthermore, as the concept of modulation might still be a bit vague for a new comer to engineering, a non-engineering example will be provided of how one could think of an equivalent to modulation in our daily lives.

In communications, the two broad categories of modulation are amplitude modulation (AM) and frequency modulation (FM) [1–4]. In both techniques, a low frequency signal is mixed with a high frequency signal. Usually, the low frequency signal is the signal of interest, either the encoding of human speech or from a similar source, and therefore can lie within a certain range of frequencies [2, 3]. The high frequency signal is usually known as the carrier signal and is tightly regulated with respect to amplitude and frequency. By mixing these two signals, the result is a signal that has some special properties thereby making it easier to transmit over long distances, and also through wireless communication. At the receiving end, the exact opposite process of demodulation is carried out, where the signal of interest is separated from the high frequency carrier signal [2, 3]. We are quite familiar with frequency modulation (FM) as all our radio channels are FM channels with different channels being of a particular frequency. As an example, depending on where we live, we could find a number of broadcasting stations using different channels to broadcast different content, and we can choose to tune our radios to a particular frequency to listen to the content.

We will not go very deep into the process of AM or FM, or for that matter examine all the different techniques possible, as the topic of communications is very vast, and not all of it is applicable to power electronics. However, we can express the basic concept of AM and FM with a few equations. Since the carrier signal is determined by us, we can start with it and express it as follows:

$$c(t) = A_c \cos(\omega_c t) \tag{2.1}$$

Where $\omega_c = 2\pi f_c$ is the angular frequency of the carrier waveform in rad/s, f_c is the frequency in Hz and A_c is the magnitude. We have chosen a simple cosine waveform as an example, though extensive research has been done on the benefits of various other waveforms. In the above equation, we can choose the frequency $f_c = 5000$ Hz to be a sample carrier frequency, though one could experiment with this value as well, and $A_c = 1$.

After choosing the carrier signal, we can now come to the signal of interest which we wish to transmit. This is the signal that we wish to modulate, and therefore, can be called the modulation signal. For the sake of an example, let us choose this signal to be a co-sinusoid as follows:

$$m(t) = A_m \cos(\omega_m t) \tag{2.2}$$

Here, $\omega_m = 2\pi f_m$ is the angular frequency of the modulation signal in rad/s, f_m is the frequency in Hz, and A_m is the magnitude. Since the modulation signal is an audio signal or a similar signal in the domain of communications, this signal will be variable both in frequency as well as in amplitude. As an example, if the audio signal was a human conversation, the signal will change in amplitude as the speakers lower or raise their voices, and will change in frequency as the pitch of the voice changes. However, to begin our discussion, we do not need to go into those details in this simple example, and can assume both the magnitude to be constant $A_m = 1$ and the frequency to be constant $f_m = 50$ Hz.

In general, one can express the result of the process of modulation as follows [1–4]:

$$y(t) = A_y \cos(\omega_y t + \theta) \tag{2.3}$$

If we wish to implement AM, the magnitude A_y will vary according to the modulation signal, while if we wish to implement FM, either the frequency ω_y or the phase angle θ will vary according to the modulation signal [1–4]. Once again, it is important to emphasize that, there are a vast number of techniques to achieve this, and numerous books and publications can be found on them. The simplest manner to achieve AM is to take the product of the modulation signal and the carrier signal, as in that case, the amplitude of the resultant waveform is directly dependent on the modulation signal.

$$y(t) = m(t)c(t) \tag{2.4}$$

As can be seen from the above equation, the frequency part of the carrier waveform is intact, but the amplitude A_y of the resultant waveform is now the modulation signal $m(t)$ itself.

Using trigonometric relations, the resultant AM waveform can be expanded as:

$$y(t) = \cos(\omega_m t) \cos(\omega_c t)$$
$$= \frac{1}{2}\cos((\omega_m + \omega_c)t) + \frac{1}{2}\cos((\omega_c - \omega_m)t) \qquad (2.5)$$

All we have used for the above simplification is the trigonometric relation:

$$\cos\alpha \ \cos\beta = \frac{1}{2}\cos(\alpha + \beta) + \frac{1}{2}\cos(\alpha - \beta) \qquad (2.6)$$

It is therefore, quite evident that the resultant waveform ends up with two combined frequencies $\omega_m + \omega_c$ and $\omega_c - \omega_m$ even though the process of AM seems to preserve the frequency of the carrier signal while using the amplitude of the modulation signal. Such modulation is merely the result of the combination of co-sinusoids or sinusoids as can be seen above. This can also be verified through a simple simulation in Python which can be found in the folder `amplitude_modulation` within the folder `chapter2` at the link: https://github.com/opensourceelectrical/switching-strategies-for-power-electronics.

To begin with, we will simulate the ideal case above, of a purely co-sinusoidal audio signal of constant frequency and amplitude being processed with AM using a carrier signal.

```python
import numpy as np
import matplotlib.pyplot as plt

# Modulation signal
fm = 50.0
omega_m = 2*np.pi*fm
# Carrier signal
fc = 5000.0
omega_c = 2*np.pi*fc

t_duration = 1.0
t_step = 1.0e-6
fsampling = 1/t_step

# Time array
time_array = np.arange(0, t_duration, t_step)
no_of_data = np.size(time_array)

# Audio signal to be modulated
mod_signal = np.cos(time_array*omega_m)

# Carrier signal
carr_signal = np.cos(time_array*omega_c)

# Resultant (modulated) signal
trans_signal = mod_signal * carr_signal

plt.figure()
plt.plot(time_array, trans_signal)

plt.figure()
plt.plot(time_array, trans_signal)
```

plt.xlim([0.78, 0.8])

plt.show()

The resultant AM waveform can be found in Figs. 2.1 and 2.2. In Fig. 2.2, it is very clear how a high frequency and a low frequency waveform have been combined by the fact that the resultant waveform is simply a high frequency waveform with a low frequency envelope. In this simulation, since we are plotting a waveform that we have synthesized, we can be sure of the fact that the resultant waveform has two frequencies—$f_c + f_m$ and $f_c - f_m$. However, in most practical cases, we are given a signal from which we have extracted samples, and with these samples we must determine which frequency components exist within them. As a result, we need to find a convenient method to perform frequency analysis on any given signal with only a collection of samples. Though at first thought, this might appear to be extremely difficult, in reality, there are readily available methods in Numpy with which we

Fig. 2.1 Result of Amplitude Modulation (AM)

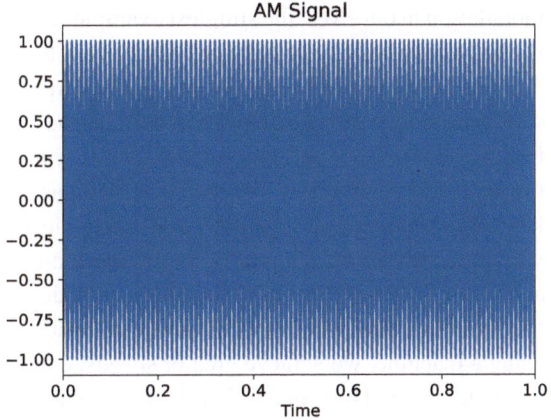

Fig. 2.2 Result of Amplitude Modulation (AM) (zoomed)

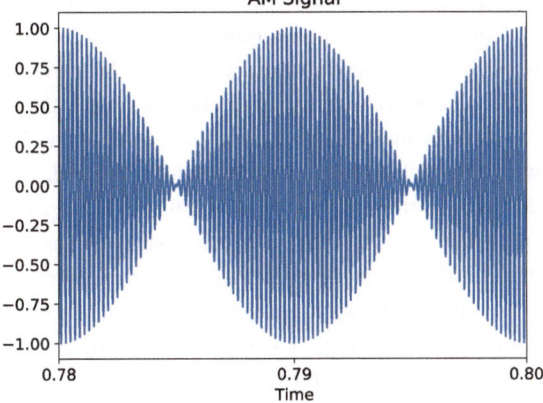

can perform frequency analysis conveniently and efficiently. This will be examined in the next section.

Before concluding this section, let us try to describe the implication of the above process of AM and why modulation is so important in communications. By performing modulation, we have simply mixed two signals - one a known and controlled signal (carrier) and the other an unknown and uncontrolled signal (modulation). In reality, all we are interested in is the modulation signal, as this is the outcome of a physical process such a human conversation over the phone. However, there are significant challenges in transmitting this modulation signal alone as we have no control over it - the frequency can be variable, the amplitude could be very low [2–4]. As a result, the process of transmission can be lossy, and at the receiving end, the waveform that is extracted could be very different from the original signal resulting in unacceptable deformation [2–4]. This is where modulation improves the process of transmission—by providing a strong and controllable signal as the base over which the signal of interest is superimposed.

If we step away from the world of communications and try to think of an analogy to modulation in our lives, the simplest example could be thought of as that of transportation. If we humans wish to move from one place to another, we have several options—walk, bicycle, drive or take a bus, take a train or even by airplane. The most basic form of movement is walking, and for short distances, would be the best, as it not only is zero cost but is also healthy. However, in terms of efficiency, walking long distances may not be optimal, as it may take too much time. Therefore, we humans would prefer to use a machine. The lowest level of a machine for movement would be a bicycle, which enables us to move much faster while still being free of fossil fuels and therefore cheap and effective. For longer distances, and when time is a constraint, we might consider driving, taking a bus or even a train. In all these cases, a human uses a machine for mobility. It is the human that wishes to move, but by using a machine, we achieve greater speeds, conserve our own energy and therefore utilize our time better. Thus, one could think of transportation as a form of modulation—we are using a faster, stronger machine which will carry us to where we wish to go. The machine is similar to the carrier signal, while we humans are the modulation signal.

Coming back to communications, it might be difficult to think of signals as strong or weak, except in terms of the amplitude. However, we could define a signal as strong, if we have full control over the frequency as well as the amplitude. As an example, when we think of radio channels, we are used to tuning our radio to different FM channels according to what we wish to hear. Each channel is a fixed, regulated frequency at which is modulated different content. Most FM channels are in the frequency range of hundreds of MHz. Using frequency modulation, it is much easier to transmit FM signals over long distances, thereby making the transmission efficient and relatively lossless, as in the end, we can listen to music, news and other content quite often with great clarity. In contrast, human speech has a low frequency range, has variable amplitude and also varying frequencies, which are very difficult to transmit on their own over long distances.

In this section, we provided a basic example through mathematical expressions and a simple simulation, the process and effect of modulation. Even though AM requires only the amplitude of the modulated signal to vary according to the modulation signal, the resultant modulated signal has different frequency components as compared to the carrier or the modulation signals. Due to this feature, there arises the need for tools to analyse the frequency components in any given waveform so as to fully understand the impact of modulation. In the next section, we will extend this simulation to describe how we can perform frequency analysis on the modulated signal.

2.3   **Introducing Frequency Analysis Techniques**

In the previous section, we examined the basic concept of amplitude modulation, and with a special case of a constant frequency modulation signal and a constant frequency carrier signal, were able to see the effect of modulation. In this section, we will introduce methods of frequency analysis [1] which we can use to determine the frequencies present in any given signal. This frequency analysis can be used throughout the book, as with the modulation techniques used with power converters that will be described in this book, the frequency components present in the converter output can be examined through frequency analysis techniques.

Though there are numerous frequency analysis techniques that one can use [1], this book will describe two very popular techniques—the Fourier series and the Fourier transform [1]. Let us start our discussion with that of Fourier series which is an easier concept to describe. The Fourier series and many forms of Fourier analysis such as the Fourier Transform were the discovery of the French mathematician Jean-Baptiste Joseph Fourier. Fourier proposed that any periodic signal $f(t)$ with a time period T_0 can be synthesized by a summation of harmonically related sinusoidal functions [1, 5]. There are many variations of the Fourier series expression, but one of the most popular ones that engineers most often use is [1, 5]:

$$f(t) = \frac{a_0}{2} + \sum_{n=1}^{\infty} \left(a_n \cos \frac{2\pi nt}{T_0} + b_n \sin \frac{2\pi nt}{T_0} \right) \qquad (2.7)$$

The above equation is called the sine-cosine form [1, 5]. a_0, a_n and b_n are coefficients that can be calculated while $\cos \frac{2\pi nt}{T_0}$ and $\sin \frac{2\pi nt}{T_0}$ are the harmonically related sinusoidal functions and are also called harmonics as they are multiples of a given fundamental pair—$\cos \frac{2\pi t}{T_0}$ and $\sin \frac{2\pi t}{T_0}$. Here, T_0 is the time period of the periodic function $f(t)$ which results in the fundamental frequency being $\omega_0 = \frac{2\pi}{T_0}$. In some cases, it will be necessary for the summation to include an infinite number of harmonics so that any given periodic function can be synthesized with such harmonic components, while in other cases, a finite number of harmonic components will be sufficient. The harmonic components can be calculated as follows [1, 5]:

$$a_0 = \frac{1}{T_0} \int_0^{T_0} f(t)dt$$

$$a_n = \frac{2}{T_0} \int_0^{T_0} f(t) \cos \frac{2\pi nt}{T_0} dt \tag{2.8}$$

$$b_n = \frac{2}{T_0} \int_0^{T_0} f(t) \sin \frac{2\pi nt}{T_0} dt$$

The great significance of the Fourier series is that given a particular periodic function, one can determine which are the dominant frequency components that compose them. As an example, if one considers a square wave with a time period of T_0, the harmonic components can be calculated using the above formula for a_0, a_n and b_n ($n = 1, 2, 3, \ldots$) with the fundamental component being $\omega_0 = \frac{2\pi}{T_0}$. For a square wave, one will need to compute an infinite number of a_n and b_n to be able to produce a series whose summation can equal the square wave. However, the calculation of a_n and b_n will give us a precise idea of how much a particular harmonic component n of the frequency ω_0 is present in a square wave. For details, the reader is encouraged to read dedicated books on signal processing which deal with the Fourier series.

An extension of the above Fourier series is the Fourier transform, that enables us to convert any function in time to a function in frequency [1], while the Fourier series was limited only to periodic functions. While the Fourier series enabled us to decompose a periodic waveform into harmonic sinusoidal components, the Fourier transform enables us to determine what would be the magnitude of any particular sinusoidal frequency component in the original function [1, 5]. Let us elaborate with describing the Fourier transform of the same function $f(t)$ above as follows [1, 5]:

$$F(\omega) = \int_{-\infty}^{\infty} f(t)e^{-j2\pi\omega t} dt \tag{2.9}$$

In the above integral, we wish to find the magnitude $F(\omega)$ of a frequency component ω in the function $f(t)$. It is important to remember that this frequency component ω may not exist and could be zero. Also, this frequency component ω has no relation to the fundamental periodic frequency $\omega_0 = \frac{2\pi}{T_0}$ of the function $f(t)$. Furthermore, while the Fourier series was applicable only to periodic waveforms, the Fourier transform can be applied to any function of time, whether it is periodic or not [1]. In the above integral, we are multiplying the function $f(t)$ at a given instant of time t with the complex sinusoid:

$$e^{-j2\pi\omega t} = \cos(2\pi\omega t) - j\sin(2\pi\omega t) \tag{2.10}$$

The integral of the product of the function at a given instant and the complex sinusoid over all time is the magnitude $F(\omega)$ of the frequency component ω in the function $f(t)$. Therefore, if $F(\omega)$ for any given frequency ω is non-zero, there will exist a frequency component [1]:

$$F(\omega)e^{j2\pi\omega t} \tag{2.11}$$

The magnitude $F(\omega)$ will be a complex value as it is the result of an integration of the product of a real-valued function $f(t)$ and a complex sinusoid $e^{-j2\pi\omega t}$.

It is very clear that the Fourier Transform is an extremely powerful technique in analysing any function of time $f(t)$ and determining the frequency components within the function. While the Fourier series is restricted to only the harmonic components of a periodic waveform, the Fourier transform can be applied to any function and can be used at any frequency of our interest [1, 5]. However, for us power electronics engineers, the above Fourier Transform cannot be directly used. Since most modern control is now digital in nature, the signals we deal with as power electronics engineers are not continuous time but rather are discrete-time. For the same function $f(t)$ above, if we were to take samples at a time interval of ΔT [1]:

$$f[n] = f(n\Delta T), \quad n = -\infty \text{ to } \infty \tag{2.12}$$

with n in the above equation being integers.

Theoretically, we can take these samples through $-\infty$ to ∞. However, in practice, we use digital controllers such as Digital Signal Processor (DSP) based micro-controllers or other equivalent devices to implement real-time control and analysis. In such devices, due to limitations of memory, there is a practical constraint on the number of samples that we can consider. Therefore, in practice, we deal with a time window during which we collect samples of the function and can perform computations on them. Therefore,

$$f[n] = f(n\Delta T), \quad n = 0, 1, 2, \ldots, N-1 \tag{2.13}$$

In the above time window, we have collected N samples of the function $f(t)$ at a time interval of ΔT.

Now that we have this digital representation of the function, we can use Discrete Fourier Trasform (DFT) instead of the Fourier Transform described above. The DFT can be expressed in a manner similar to the Fourier Transform, except that the integral is replaced with a summation and the function $f(t)$ is replaced with the sampled function $f[n]$ [1, 5]:

$$F[k] = \sum_{n=0}^{N-1} f[n]e^{-j2\pi\frac{kn}{N}} \quad ; k = 0, 1, \ldots, N-1 \tag{2.14}$$

In the above example, N is the total number of samples of the function within the time window under consideration. The expression $e^{-j2\pi\frac{kn}{N}}$ is merely a mathematical form of writing the complex sinusoidal relationship:

$$e^{-j2\pi \frac{kn}{N}} = \cos\left(2\pi \frac{kn}{N}\right) - j \sin\left(2\pi \frac{kn}{N}\right) \tag{2.15}$$

Therefore, every sample $f[n]$ in the N-sample window is being multiplied with the above complex sinusoidal relationship and the sum total over all N samples in the window is then being assigned to a particular sample k of the frequency component $F[k]$. The only question arises is what is this sample k? Since we are dealing with N samples of the function f, we can calculate N samples of the frequency component F. Therefore, while the subscript n refers to any sample of the function in the N-sample window, the subscript k refers to any sample of the frequency component in the N-sample window. This concept can be a bit confusing. The simplest approach to understanding this is as follows. The continuous-time Fourier Transform will integrate over a continuous-time function $f(t)$ over all time instants from $-\infty$ to ∞ to produce frequency components $F(\omega)$ over all frequency from $-\infty$ to ∞. On the other hand, the DFT will sum over N samples of a function f to produce N samples of frequency components F.

There is one major drawback to sampling a particular signal. Once we sample a signal, and extract the samples into an array as has been done in the simulation and in any real-time control or analysis, we have in principle lost the knowledge of time. In reality, we can always find out the exact time instant of a sample by multiplying the sample n with the time interval ΔT. However, from a mathematical perspective, we only consider the samples as a series of values and not at particular instants of time. When performing DFT, one can notice how the sampling time interval ΔT is completely missing and only the total number of samples N appears. This is a case of how when performing a transformation on a sampled signal, one uses only a particular sample n or k and the total number of samples N. The resultant of the DFT $F[k]$ is a series of samples of frequency and this resultant also has the same number of samples N as the sampled signal $f[n]$ [1, 5]. Therefore, for a time window of samples of a function, we now have a frequency window with the same number of samples.

The DFT is similar to the Fourier Transform in that the resultant samples $F[k]$ are magnitudes of the frequency components in the sampled function $f[n]$ [1, 5]. However, what is puzzling is what the samples represent. In the case of the Fourier Transform, the result of the transformation $F(\omega)$ can be computed for a particular frequency ω. However, for the DFT, the resultant is merely a sample $F[k]$ over a window with the same number of samples. To make sense out of what DFT produces, we need to deviate from the mathematical ideal and try to figure out as engineers, how we are going to use this sample of frequencies. It is important to note that the following explanation is not strictly mathematically rigorous, but is merely an interpretation that helps us to make use of the DFT. Luckily, we have helper functions with Numpy to perform all this with code, thereby making our lives much simpler [6–8].

The key here is to take into consideration the sampling time interval ΔT and therefore, the corresponding sampling frequency. In the simulation, the sampling time interval ΔT is 1 μs and therefore, the sampling frequency is 1 MHz. In such a case, the range of frequencies in the output of the DFT will span this sampling frequency of 1 MHz. However, due to the nature of DFT, the frequency samples contain both negative and positive values. This might seem confusing at first. Within the DFT, there are many variations [1, 5]. One can compute the DFT of a sampled signal only for positive frequencies between 0 to $N - 1$ [1, 5]. One can also compute the DFT of a sampled signal between the frequency range $-\frac{N}{2}$ to $\frac{N}{2} - 1$ [1, 5]. The `fft` method that we are using in the simulation which is provided by the `fft` package within Numpy provides the DFT between the frequency range $-\frac{N}{2}$ to $\frac{N}{2} - 1$. Therefore, we need to take this into account while interpreting the frequency of the samples produced by DFT. We do this by using the `fftfreq` method in the `fft` package provided by Numpy. This method takes in the number of samples and the sampling time interval to produce an array of frequencies ranging from $-\frac{2\pi}{\Delta T}$ to $\frac{2\pi}{\Delta T}$.

The code for performing DFT using Python are as follows. These can be inserted into the above code as necessary:

```
from numpy.fft import fft, fftfreq

# FFT of the entire waveform
freq_trans_signal = fft(trans_signal)/no_of_data
freq_array = fftfreq(no_of_data, t_step)

plt.figure()
plt.plot(freq_array, np.abs(freq_trans_signal))
plt.xlim([-50000, 50000])

plt.figure()
plt.plot(freq_array, np.abs(freq_trans_signal))
plt.xlim([4800, 5200])

plt.show()
```

Theoretically, the sampled signal $f[n]$ can be complex. As power electronics engineers, the signals we deal with are always real, but mathematically, DFT can be applied to any signal real or complex [1]. The output of the DFT is also a series of complex values with real and imaginary parts due to the multiplication of every sample $f[n]$ with the complex sinusoidal exponent $e^{-j2\pi \frac{kn}{N}}$ [1]. However, for us power electronics engineers, we are only concerned with the total result, and therefore, we can consider the absolute values of the complex quantities. Additionally, due to the summation over N samples that occurs with the DFT, it is necessary to normalize the result, due to which the output of the `fft` method has been divided by the total number of samples N. As a result of this division by the number of samples, the frequency components appear as real absolute values which is very convenient for the purpose of frequency analysis.

With this background, we can analyse the output of the DFT using Figs. 2.3 and 2.4. Figure 2.3 shows the entire range of frequencies from –500 kHz to 500 kHz. Figure 2.4 shows the frequency range around 5000 Hz which is the frequency of the carrier waveform. It is evident from Fig. 2.4 that the modulated waveform contains the two side bands $f_c - f_m$ and $f_c + f_m$. If one zooms in on the plot in the 50 Hz frequency range, one will find no frequency component, which indicates that modulation has pushed the frequency of

Fig. 2.3 Frequency response of
amplitude modulated signal

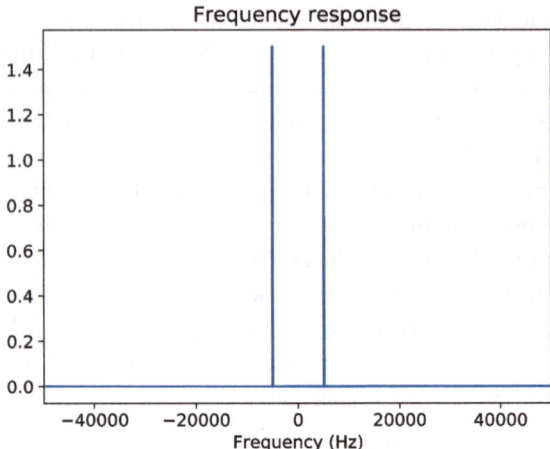

Fig. 2.4 Frequency response of
modulated signal (zoomed)

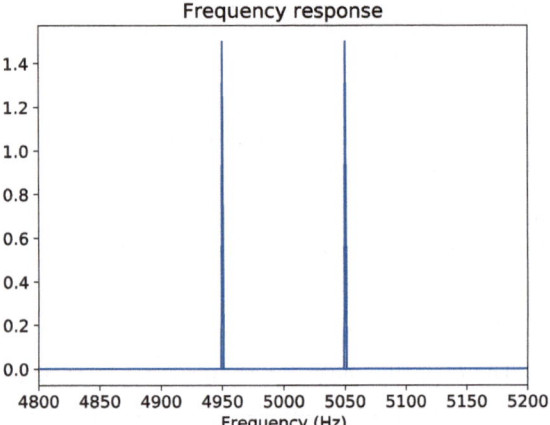

the modulation signal farther away towards the carrier signal frequency. This verifies our
equation for the modulated waveform. Moreover, one can observe that the magnitude of the
side bands is 0.25. However, if one recollects that there are also side bands $-f_c - f_m$ and
$-f_c + f_m$ in the negative frequency range, and these side bands are also having a magnitude
of 0.25, then it is clear that the absolute frequency side bands $|f_c - f_m|$ and $|f_c + f_m|$ have
a magnitude of 0.5 each. The result of AM in our simple case is:

$$\frac{1}{2} \cos((\omega_m + \omega_c)t) + \frac{1}{2} \cos((\omega_c - \omega_m)t) \tag{2.16}$$

Therefore, the DFT verifies that the frequency side bands have an equal magnitude of 0.5
and form the entire modulated signal. The reader is encouraged to change the magnitude of
the modulation signal and examine the result of the DFT.

Now that we have presented a technique for frequency analysis, let us examine how this analytical tool behaves if the modulation signal had a certain degree of randomness, rather than being a constant frequency and constant amplitude signal as considered before. This would be the more practical case when the modulation signal is an audio signal whose amplitude and frequency can vary if it were a recorded voice conversation. Let us consider a modulation signal:

$$m(t) = (A_m - r_x(t)) \cos((\omega_m - r_y(t))t) \tag{2.17}$$

Where $r_x(t)$ and $r_y(t)$ are random functions. Random numbers can be generated using the random method within the random package in Numpy. This random method produces a floating point number between 0 and 1. We can use a scaling factor to scale it to a desired value for our application. As an example, to make the amplitude deviate by 0.3 and the frequency by 30 Hz, we could use the following statement:

from numpy.random **import** random as rm

```
mod_signal = (Am − rm()*0.3) * \
             np.cos(time_array*(omega_m − 2*30.0*np.pi*rm()))
```

The Python code for the example can be found in the folder amplitude_modulation within the folder chapter2 at the link: https://github.com/opensourceelectrical/switching-strategies-for-power-electronics.

The results can be described in Figs. 2.5 and 2.6. Figure 2.6 shows that the side bands have changed from before and are no longer centered around 4950 Hz ($f_c - f_m$) and 5050 Hz ($f_c + f_m$). This is due to the fact that the modulation signal has dominant components with a frequency lesser than 50 Hz. Furthermore, it can also be observed that the side bands are not a single dominant frequency as in the previous case, but a number of frequency components can be found in the side bands. Therefore, the frequency analysis has captured all the frequency components in the resultant modulated signal. The reader is encouraged

Fig. 2.5 Frequency response of amplitude modulated signal

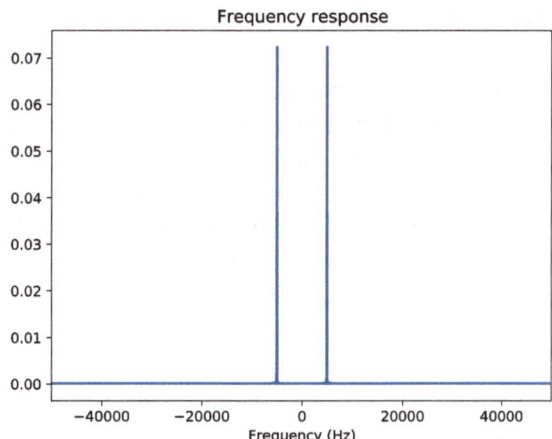

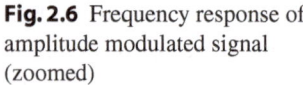

Fig. 2.6 Frequency response of amplitude modulated signal (zoomed)

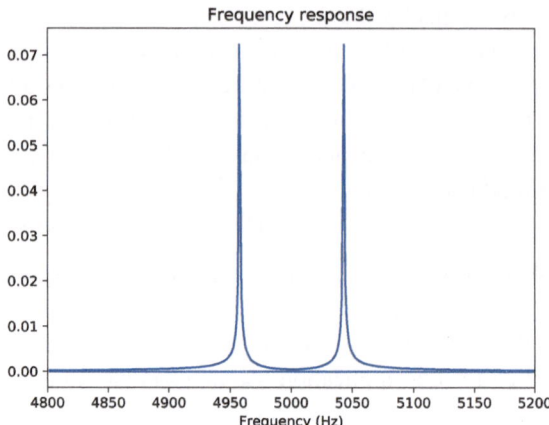

to print out the frequency components and corresponding magnitudes of the modulation signal, and verify whether these components have been captured in the frequency spectrum.

This section described two very powerful and popular frequency analysis techniques, namely the Fourier series and the Fourier transform, that are used extensively by power electronics engineers. Many commercial scientific and simulation software provide these tools as ready functions that can be used conveniently. However, to fully understand how these tools can be used, one needs a basic understanding of theory. A reader might assume that one might have to implement these frequency analysis techniques from scratch if using a free and open-source programming language such as Python. However, since Python has a vibrant scientific community, many frequency analysis tools are available with both Numpy and Scipy. The section described a convenient manner in which DFT can be performed on any signal that is available as a collection of samples.

2.4　Frequency Modulation (FM)

In the previous sections, we described in brief the concept of AM and how it can result in a transformation of frequencies in the resultant waveform. Frequency Modulation (FM) behaves in a similar manner, except that we are directly regulating the frequency of the modulated waveform with the modulation signal, unlike the case of AM, where we were varying the magnitude of the modulated signal directly, though this indirectly ended up varying the frequency of the modulated signal. FM is preferred over AM as it is possible to achieve higher frequency ranges, and the infrastructure needed to transmit an FM signal is cheaper and more efficient than an AM signal [2–4]. These details are not necessary in this book, as these concepts of modulation are merely a stepping stone to the modulation used in power electronics. However, the interested reader can find plenty of literature on modulation techniques used in communications [1–4].

Repeating the general equation for a modulated waveform:

$$y(t) = A_y \cos(\omega_y t + \theta) \tag{2.18}$$

In FM, we will vary the frequency ω_y with respect to the modulation signal. As with AM, there are numerous techniques available to make this possible [2–4]. Let us choose the simplest conceptual representation of [1]:

$$y(t) = A_y \cos((\omega_{y0} + A_m \cos \omega_m t)t + \theta) \tag{2.19}$$

The only change that needs to be made in the code for AM is the one statement that deals with the modulated signal synthesis. In addition, we can skip the waveform plot of the modulated waveform and only examine the frequency spectrum.

```python
import numpy as np
import matplotlib
import matplotlib.pyplot as plt
from numpy.fft import fft, fftfreq

fm = 50.0      # frequency of audio signal
omega_m = 2*np.pi*fm
fc = 10000.0      # frequency of carrier signal
omega_c = 2*np.pi*fc

t_duration = 1.0
t_step = 1.0e-6
fsampling = 1/t_step

# Time array which is integers from 0 to 1 million - 1
time_array = np.arange(0, t_duration, t_step)
no_of_data = np.size(time_array)

# Audio signal to be modulated
mod_signal = np.cos(time_array*omega_m)

# Carrier signal
carr_signal = np.cos(time_array*omega_c)

# Resultant (modulated) signal
trans_signal = np.cos(time_array*(omega_c + mod_signal))

# FFT of the entire waveform
freq_trans_signal = fft(trans_signal)/no_of_data
freq_array = fftfreq(no_of_data, t_step)

# Plot of the frequency spectrum modulated signal
plt.figure()
plt.plot(freq_array, np.abs(freq_trans_signal))
plt.title('Frequency response')
plt.xlim([-50000, 50000])

plt.figure()
```

```
plt.plot(freq_array, np.abs(freq_trans_signal))
plt.title('Frequency response')
plt.xlim([9800, 10200])

plt.show()
```

The Python code for the example can be found in the folder `frequency_modulation` within the folder `chapter2` at the link: https://github.com/opensourceelectrical/switching-strategies-for-power-electronics.

The waveform of the resultant modulated waveform will have a constant magnitude as only the frequency is being varied. Since the frequency of the modulation signal is a fraction of the carrier signal, the changes in frequency of the modulated signal will be almost imperceptible. Therefore, a zoomed in waveform of the modulated signal has not been shown, though the reader is encouraged to try out the simulation and zoom in on the plot of the modulated waveform. What is of great interest is the frequency spectrum of the modulated signal as shown in Figs. 2.7 and 2.8. From Fig. 2.8, the largest component is the base carrier frequency (10 kHz), followed by several side bands that are separated from each other by the frequency of the modulation signal (50 Hz). Unlike the case of AM, where a single side band existed, in FM, the side bands progressively increase on either side of the carrier frequency, with the magnitude of the side bands continuously decreasing until their magnitudes become negligible [1]. The reader can confirm that similar to the AM, there are no frequency components in the frequency range of the modulation signal (50 Hz), and all the frequency components have been moved to be close to the frequency of the carrier signal.

At first glance, FM might appear to have produced a much more complicated resultant waveform with several frequency side bands. In comparison, AM produced a fairly well-behaved modulated waveform, where the frequency components present could be predicted without too much fuss [2–4]. At this point, it needs to be stated, that many well-established

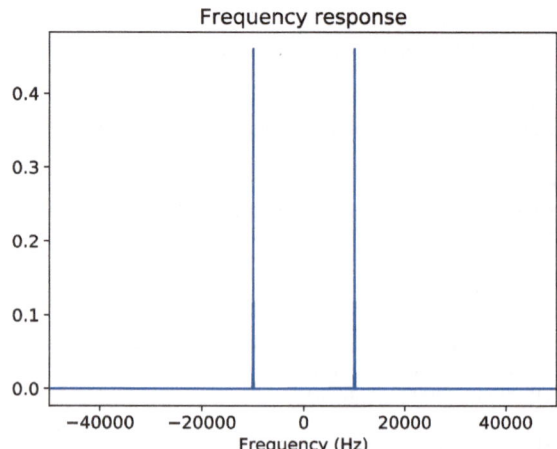

Fig. 2.7 Frequency response of frequency modulated (FM) signal

Fig. 2.8 Frequency response of frequency modulated (FM) signal (zoomed)

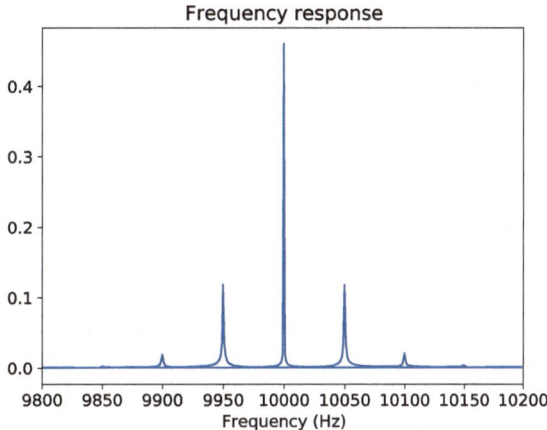

techniques are available to extract the modulation signal from the resultant modulated wave-form, both for AM as well as FM. Since this book is not about communications, there is little point in discussing them in detail. However, the prime reason for presenting the case of FM with a sample simulation is to show the frequency spectrum and some of the special characteristics of it, as these will be encountered in the next section, where we begin with modulation in the domain of power electronics.

2.5   A Basic Overview of Power Electronic Applications

In the previous sections, we had described how modulation is used in the domain of communications. The reason for doing so was partly because as electrical engineers, a basic course on signal processing usually precedes a course on power electronics. Moreover, with all of us exposed to a vast number of digital communication devices, a certain basic understanding of communications is almost intuitive. However, though some concepts of modulation can be directly translated from communications to power electronics, there are significant differences, primarily due to the fact that applications in power electronics are typically high (or comparatively higher power) and the main objective is to transform electrical energy from one form to another. In this section, we will gradually introduce these differences in power electronic applications, and subsequently introduce the use of modulation.

Though the applications of power electronics are vast, if one were to describe power electronics in a single line, it would be those applications of electronics that enable us to convert electrical energy from one form to another. When we speak of different forms of electrical energy, we can speak of voltage levels, frequencies or even the number of phases. Let us examine a few examples that everyone is familiar with. A mobile phone charger converts the voltage available from an ac outlet into 5 V dc that can be fed to the mobile phone battery through a USB cable. The ac outlet can be 240 V, 50 Hz or 120 V, 60 Hz or some other variation of these depending on the geographic location. Therefore, we are converting an ac voltage of a particular voltage level and frequency into a dc voltage (zero frequency) and 5 V. There is a conversion in both voltage level as well as frequency. What is remarkable is that nowadays most mobile chargers are small enough to fit into one's pocket.

In the case of a solar photovoltaic (PV) panel connected to the grid and supplying clean energy into our power system, the conversion is almost the opposite. The PV panel will produce a dc voltage that varies according to the intensity of incident solar radiation. A solar inverter is used to interface this variable dc voltage and inject an ac current into the power grid. This ac current will have a fixed frequency equal to that of the grid frequency and must also have a phase relationship with respect to the grid voltage such that it injects power into the grid. Due to the variable nature of the dc voltage from the PV panel, the current injected into the grid will have a variable magnitude. In this case, we have transformed a dc voltage into an ac current of fixed frequency. It is due to the increasingly sophisticated nature of these solar inverters, that we are able to meet a greater percentage of our energy needs from green energy in our fight against climate change.

Let us consider one more case—that of an electric vehicle (EV). In an electric vehicle, we use a battery as an energy source to supply a motor. Battery technology is advancing at a rapid pace, and it is beyond the scope of this book to describe accurately the type of batteries used in modern EVs. However, one of the most popular EV battery is the Lithium-ion (Li-ion) battery which is a high density battery that enables even large sized EVs to have a range of a few hundred kilometres on a full charge. The battery produces a dc voltage whose voltage is at the highest when the battery is fully charged, and this voltage decreases as the battery charge decreases. The motor on the other hand is an ac motor, either an induction motor or a synchronous motor. Furthermore, for a small EV, one could consider using a single-phase motor, while for a large-sized car, a three-phase motor will need to be used. An inverter performs the task of converting the dc voltage produced by the battery into an ac voltage which could be potentially multi-phased to supply the ac motor.

The above three cases were just a few sample cases as these are easily understandable for a novice electrical engineer. However, the above cases are a very convenient starting point to describe the basic principles behind a power electronic converter. Let us choose the case of converting a given dc voltage into an ac voltage of a fixed frequency which for the sake of an example we can choose as 50 Hz. Let us for a moment not think of a power electronic solution to this problem, but let us think of any random solution. A quick solution that jumps to the mind of an electrical engineering student would be to use a coupled motor-generator set, as such a coupled set is quite often used in demonstrating basic experiments for junior undergraduate students. The dc voltage could be fed to a dc motor which is mechanically coupled to an ac generator. Therefore, by supplying a dc voltage to the dc motor, we are causing it to rotate, thereby providing mechanical input to the ac generator. The ac generator will produce an ac voltage.

This accomplishes one part of the problem—conversion of dc voltage to ac voltage. However, we need to produce an ac voltage of a particular frequency. To achieve this, we need to enable speed control in the dc motor such that for any given dc voltage supplied to it, the speed of the dc motor will remain constant. Due to the constant speed for the dc motor and therefore the ac generator, the frequency of the ac voltage produced will now be constant. If we wish to further regulate the magnitude of the ac voltage, we need to implement excitation control in the ac generator such that the flux in the ac generator is of such a magnitude that the output ac voltage is of a constant magnitude. As undergraduate students, we have many courses dedicated to motors and generators, where one has many such coupled motor-generator sets with which a number of different scenarios can be emulated.

Before the advent of power electronics, motors and generators were the only solution to achieving power in a particular form. Nowadays, power supplies and inverters are commonly used to supply sensitive electronic devices to ensure that they receive stable and continuous power. A few decades before, diesel generators were commonly used as backup power generators to supply systems that were critical. As an example, many high-rise buildings would have backup diesel generators to enable elevators to continue functioning in the event of a power failure to ensure that the occupants of a given building could leave the building safely and not be stranded inside. Nowadays, many of these diesel generators are being replaced with inverters with battery backup, as these inverters and batteries need less maintenance than diesel generators. The question then can be asked—what makes an inverter have less maintenance than the conventional rotational machines?

Power electronics similar to all other conventional electronics has one characteristic feature—these are static systems with no moving parts. As an example, a mobile charger or a solar inverter can convert dc voltage into ac voltage or current without any rotational or moving parts. In the case of the EV, though the application demands motion, the inverter itself is static and has no moving parts. Only the motor which is being supplied by the inverter rotates and produces the desired movement. A static system has an immediate advantage over a moving system, as the absence of motion avoids several problems such as no vibration, simplified cooling, and no friction and associated wear-and-tear that needs lubrication. In most cases, no or less movement automatically translates to fewer headaches.

Power electronics has a very distinctive difference with respect to conventional analog electronics. As an example, in analog electronics, one of the most common applications is that of an amplifier—when we have a weak signal that we wish to magnify so that it can be made use of. This is usually achieved using a transistor such as a Bipolar Junction Transistor (BJT) or a Field Effect Transistor (FET). There are numerous amplifier circuits, however, for the sake of an example, let us consider a common emitter configuration. In such a configuration, the input weak signal is connected to between either the base and emitter of a BJT or between the gate and source of a FET while the output is extracted between the collector and emitter of a BJT or the drain and source of a FET. Without going into details, the transistor typically works in the active region with an amplification equal to the forward current gain which is a device characteristic.

In the case of consumer electronics, the power levels being low, transistors can be operated in the active region such as with the case of the amplifier above. This is due to the fact that the losses in the transistor will be equal to the product of the voltage across the collector and emitter in the case of a BJT or across the drain and source in the case of a FET, and the collector current in the case of a BJT or the drain current in the case of a FET. In the active region both the collector-emitter (or drain-source) voltage and the collector current (or the drain current) of the transistor are non-negligible. In the cut-off region, the collector current (or the drain current) is negligible, while in the saturation region, the collector-emitter voltage (or the drain-source voltage) is negligible, resulting in significantly lower losses. However, analog amplifiers typically work in the active region and not in the cut-off or saturation regions, as at low power levels, the losses in the active region are acceptable.

In power electronic applications, with the power level being significantly higher, the corresponding power transistors cannot work in the active region as the power losses will be too high and will damage the device. Power transistors or other power devices work either in the cut-off region or the saturation region where the power losses are significantly lower and can be dissipated safely. Though this might seem like a convenient solution, it presents a complexity. In the case of analog transistors working in the active region, one could use the transistor forward current gain for designing the circuit. This forward current gain is a device characteristic and though can vary under certain operating conditions, it is fairly predictable, and therefore, can result in reasonably well-designed linear circuits. Since power devices do not work in the active region, we cannot use a device characteristic to conveniently design power electronic circuits to generate a particular output given a particular input.

When a power device is in the cut-off region, it conducts a negligible current and therefore, behaves as an open switch. On the other hand, when a power device is in the saturation region, the voltage drop across it is negligible and therefore, behaves as a closed switch. Since the power device can operate in only these two regions, we are therefore constrained to these two extreme cases—one where it behaves as an open switch and one where it behaves as a closed switch. With these constraints, we need to achieve our goals as mentioned above—for example, converting a dc voltage produced by a solar PV panel into an ac current. It is quite obvious, that this constraint has significantly complicated our problem. This is what makes power electronics extremely challenging and interesting—the need to design systems that can achieve high power equivalents of consumer analog circuits but without the freedom to operate in the active region.

To give an additional insight into why power electronics is complicated, let us compare it with the domain of switchgear and protection. In the domain of electrical protection, we use devices such as relays and circuit breakers to disconnect parts of a circuit that might be faulty, and to reconnect them back to the rest of the circuit when the fault has been cleared. Relays and circuit breakers also have usually two states of operation—one where they behave like a closed switch and the other where the behave like an open switch. However, when we wish to protect an electrical system by disconnecting a faulty part, this is achieved by breaking the connection between the faulty part and the rest of the system. Therefore, though a circuit breaker has only two operating states, the functionality that it needs to provide is also only dual—disconnection or connection. In contrast, in power electronics, with only a two state operation of power devices, a fairly complicated goal needs to be accomplished. As an example, to control an EV, the inverter has to provide adequate voltage to the motor such that it generates the torque needed for acceleration or deceleration, with the power devices in the inverter only allowed to operate in the cut-off or saturation regions.

In this section, we examined a very basic overview of power electronic applications and how they differ from conventional magnetic machines and also from consumer low power electronics. With this background, in the next section, we will examine using some very basic conceptual simulations, how power electronics is able to achieve these goals, and why it becomes necessary to use modulation techniques.

2.6   Introducing Modulation in Power Electronics

In the previous section, we had presented a very abstract overview of power electronic applications, and what makes them different. The previous section had avoided details, but rather tried to present the challenge of power electronics in an extremely simple and understandable manner. To repeat the most important argument which the reader needs to remember while reading this section—how can we achieve complex conversions of energy with respect to frequency, voltage level and many more while having at our disposal only devices that behave like switches. The reason for this constraint as already been said is due to the power losses in the devices becoming unmanageable if a third state besides the on and off state were to be introduced. For complex applications, such as high frequency converters, it might become necessary to add additional constraints to minimize switching losses. However, these advanced applications will not be considered in this book.

For the moment, let us not consider power devices, but let us consider only switches. Furthermore, let us not consider a state where the switch is supplying current, but instead let us consider how we can use only the conduction state of a switch to synthesize other waveforms. For the reader to be able to follow this section, he or she must be familiar with the concepts of frequency analysis presented in the previous sections along with the Python simulations as this section will extensively use Python simulations to prove concepts. For the sake of an example, let us suppose we want the switch to follow the pattern of a 50 Hz sine waveform. Since this is merely an illustrative example, the magnitude of the sine waveform is not important and therefore, the magnitude of the sine wave can be set at unity or 1. Moreover, the state of the switch can be thought of as either ON (1) or OFF (−1). In most digital applications, a binary digit (or bit) is considered to be 1 or 0. We have chosen the states to be 1 or −1, as we are trying to synthesize an ac waveform, which would not be possible with only 1 and 0.

The following code segment can be used to generate the switching pulses for the switch to follow a sine waveform:

```
import numpy as np
import matplotlib.pyplot as plt
```

```
def compare(a):
    if a > 0:
        return 1
    else:
        return −1
vcompare = np.vectorize(compare)

fsqw = 50.0
omega_sqw = 2*np.pi*fsqw
t_duration = 1.0
t_step = 1.0e−6
fsampling = 1/t_step

time_array = np.arange(0, t_duration, t_step)
no_of_data = np.size(time_array)

# Sine wave
sine_wave = np.cos(omega_sqw*time_array)

# Square wave signal
sq_wave = vcompare(sine_wave)

# Plot of the square wave signal
plt.figure()
plt.plot(time_array, sine_wave)
plt.plot(time_array, sq_wave)
plt.title('Voltage Output')
plt.xlim([0.0, 1.0])
plt.xlabel('Time')

plt.figure()
plt.plot(time_array, sine_wave)
plt.plot(time_array, sq_wave)
plt.title('Voltage Output')
plt.xlim([0.775, 0.805])
plt.xlabel('Time')
plt.xticks([0.78, 0.79, 0.8])

plt.show()
```

The Python code for the example can be found in folder PWM within the folder chapter2 at the link: https://github.com/opensourceelectrical/switching-strategies-for-power-electronics.

Figure 2.9 shows a sine wave of unity magnitude. It might seem impossible to synthesize such a waveform with only two levels. In the code sample above, we get as close as possible, and that would be with a rectangular waveform with a magnitude +1 during the positive half-cycle of the sine wave and magnitude −1 during the negative half-cycle of the sine wave. If the reader recollects the discussion on Fourier series in the section of frequency analysis, any periodic waveform can be synthesized using harmonics of sine and cosine waveforms. Therefore, if the sine waveform had a time period of T_0, the rectangular waveform produced

Fig. 2.9 Rectangular waveform and sine waveform of same frequency

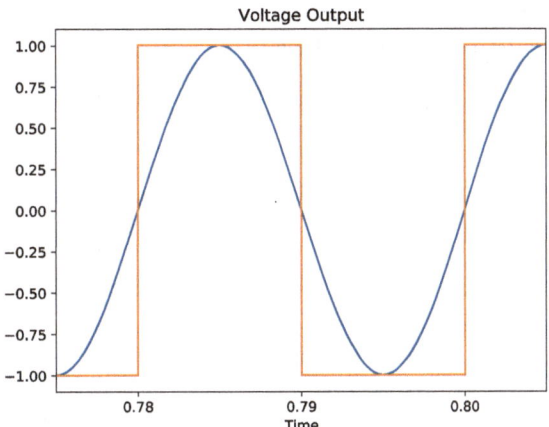

by the switch would also have a time period of T_0. Therefore, this rectangular waveform can be represented as the following Fourier series:

$$f(t) = \frac{a_0}{2} + \sum_{n=1}^{\infty} \left(a_n \cos \frac{2\pi nt}{T_0} + b_n \sin \frac{2\pi nt}{T_0} \right) \tag{2.20}$$

where,

$$f(t) = +1; \quad kT_0 \le t < (k+1)\frac{T_0}{2}$$

$$= -1; \quad (k+1)\frac{T_0}{2} \le t < (k+1)T_0$$

and the coefficients a_0, a_n and b_n are expressed in (2.8).

One could calculate a_0, a_1, b_1, a_2, b_2, a_3, b_3, and so on, and as a greater number of harmonics are considered, one can verify how close the synthesized waveform is to the rectangular waveform. Such a method is a brute force method of calculating the harmonic components. However, as already discussed in the section of frequency analysis, one can use the Fourier Transform to analyse the magnitude of a particular frequency component in the rectangular waveform. In our simulation, we can perform a Discrete Fourier Transform (DFT) of the samples in the rectangular waveform to generate the frequency spectrum. Since the rectangular waveform in our case is a periodic waveform, the frequency spectrum will merely contain the harmonic components a_0, a_n and b_n presented above.

Figure 2.10 shows the complete frequency response while Fig. 2.11 shows the frequency spectrum in the range from 0 to 1000 Hz. Though the generic equation of the Fourier series contains an a_0 component which is a dc (or zero frequency) component, the frequency spectrum does not have any component at 0 Hz. Such a dc component occurs when the periodic waveform has a dc shift, which in our simulation is not the case. The prime frequency component is the fundamental 50 Hz (or $\frac{1}{T_0}$) component which can be computed from a_1,

Fig. 2.10 Frequency spectrum of the rectangular waveform

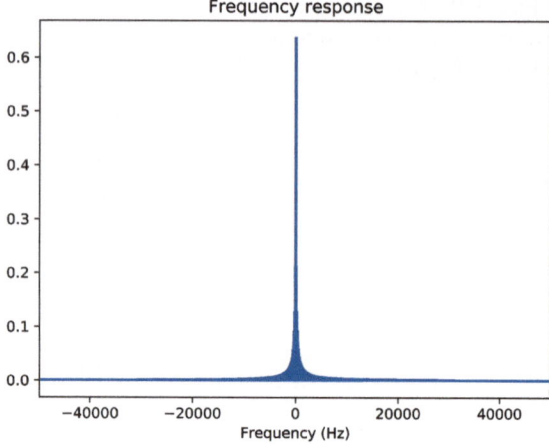

Fig. 2.11 Frequency spectrum of the rectangular waveform (zoomed)

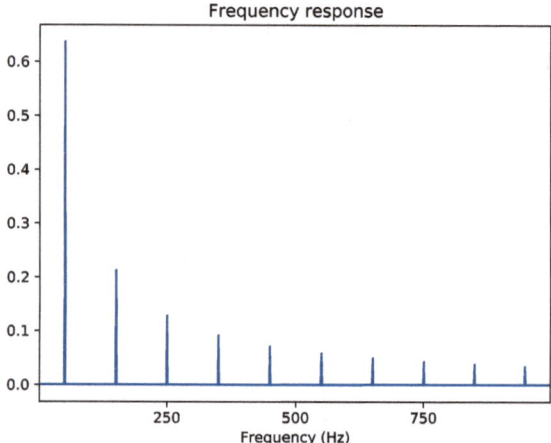

b_1. Besides these fundamental harmonic components, one can see harmonic components having frequencies 150, 250, 350, 450 Hz, and so on. Figure 2.10 shows the large spread of the harmonics. One can quickly notice that these are the odd harmonic components, while the even harmonic components 100, 200, 300 Hz, and so on are absent. To examine in detail why even harmonics are absent while odd harmonics are present is not necessary at this stage, though the reader can attempt to solve the above integral equations for a_n and b_n for odd and even values of n and arrive at that conclusion. A few other conclusions are however, much more important at this stage.

Since we are performing a frequency analysis on a rectangular waveform, the spectrum of harmonic components present is almost infinite. This is due to the rectangular nature of the waveform, which will need an infinite number of sinusoids to be able to synthesize it. However, as the frequency of the components increases, the magnitude of the components decreases until they are almost negligible. Therefore, their presence is merely a mathematical fact, whereas from the engineering perspective, they almost do not exist. With these few facts established, let us ask the question—what needs to be done now? Our intention was to synthesize a sine waveform using a switch that can achieve only two states. We are now seeing from the frequency spectrum that though the rectangular waveform produced by the switch contains the sine waveform of 50 Hz that we are interested in, it also contains a number of harmonics that we have no need for. Since our requirement was only for a 50 Hz sine waveform, we must find a way to remove all the other harmonics.

In signal processing, when we wish to retain a frequency component, and eradicate other frequency components, we need to design a filter. Filters can be of various types—low pass filters, band pass filters, band stop or notch filters, high pass filters and a few other advanced filters. A low pass filter will allow low frequency components to pass through and will block higher frequency components. A band pass filter will allow a certain band of frequency components that are close together to pass through and will block all frequency components outside this band—either having frequencies lower or higher than the range of frequencies in the band. A band stop filter will block all frequency components within a band of frequencies close together and will allow all other frequency components to pass through. A high pass filter will allow high frequency components to pass through and will block low frequency components.

If we wish to retain the 50 Hz component and block all other frequency components present, we could either design a low pass filter or a band pass filter. A low pass filter will be acceptable as there is no frequency component lower than the 50 Hz component in the frequency spectrum and therefore, if we set a cut-off frequency of around 80–100 Hz, we could let the 50 Hz component pass through and block all the harmonics. A band pass filter could also work if we choose the band of frequencies that we wish to pass as between around 40–60 Hz. Therefore, the 50 Hz component lying within this band will pass through while the harmonics that have a frequency higher than the band will be blocked. Though this can be achieved theoretically, in practice, this particular task of filtering is quite messy and will result in the filter components being bulky as the cut-off frequencies are quite low.

When one wants to design a filter, two questions need to be asked—where is the frequency of interest, and how close is the first nuisance frequency? In our case, the frequency of interest is 50 Hz, while the closest nuisance frequency is 150 Hz. The 100 Hz frequency difference is considered to be dangerously close for the purpose of design. It could be considered similar to shooting at a target while not wishing to pose a danger to anything else that might be close to the target. If there is another object close to the target, then a shooter might decide not to take the shot out of fear of hitting the other object. Of course, filters can be designed for almost any application, and a solution exists for such a case as well. However, to be able to design a filter that can precisely differentiate between frequency components close to each other will need a great deal of effort and also expensive components. Moreover, as the filter components age, there is always the danger that the harmonics will end up creeping into the resultant waveform leading to distortions.

With this discussion, we can bring in our previous discussion of modulation in communication. In communication, we wished to transmit a weak low frequency signal without distortions, and found that rather than attempt to transmit the low frequency signal, it would be better to mix it with a high frequency carrier. What if we perform the same action in power electronics as well? We wish to achieve a 50 Hz sine waveform. Instead of trying to produce it with a low frequency rectangular waveform, what if we find a way to mix the low frequency rectangular waveform with a high frequency carrier signal? In communication, we saw the resultant waveform had frequencies only in the high frequency range with frequencies in the modulation signal appearing as side bands around the high frequency carrier signal. The only question we must ask as power electronics engineers is—can we achieve such a frequency shift while still being able to extract our fundamental frequency component using simple filters? This last question is important, as in communications, the process of extracting the modulation signal from the modulated waveform called as demodulation, can sometimes be fairly complicated.

In this section, we have examined at a very basic conceptual level, how with a simple switch which can achieve only two states, it is possible to achieve a range of frequencies including the sine waveform of the fundamental frequency that we are interested in. Therefore, we have partially achieved our goal. The obstacle is in removing all the other unwanted harmonic components from the resultant waveform. Due to the insufficient separation between the fundamental component and the harmonic components, designing a filter will be difficult and expensive. In the next section, we will examine how modulation can be used in controlling this same switch to solve the second part of our problem.

2.7   Pulse Width Modulation (PWM)

In the previous section, we had described how with only a switch that can attain two states (+1 and −1), it is theoretically possible to synthesize a sine waveform. This strategy has one obvious advantage which is of ease of control in terms of regulating the state of the switch. However, in order to produce an acceptable final sine waveform, one needs to design a fairly expensive filter to remove all the other components that are present in the rectangular waveform but are not needed. With this explanation of the need for some form of modulation in the synthesis of the final waveform, in this section, we will examine the strategy of Pulse Width Modulation (PWM) which is almost ubiquitous in power electronics, though there are several strategies to achieve PWM [9–12].

Repeating the basic definition of modulation—a low frequency signal is mixed with a high frequency signal in order to produce a waveform whose frequency spectrum has certain properties that make it convenient to deal with. In the example of the switch that can be either +1 or −1, the result of the modulation has to control the state of the switch. Therefore, whatever we decide to do in terms of modulation, the end result is to regulate the state of the switch to be either +1 or −1. In the previous section, we had arrived at this by simply stating that when the modulation signal (the sine wave) was greater than 0, the switch status would be +1, and when the modulation signal would be lesser than 0, the switch status would be −1. Therefore, we had used the modulation signal to directly regulate the state of the switch. In this section, we are going to use this modulation signal in the process of modulation to create another signal which will regulate the state of the switch. Since we are eventually regulating the state of the switch, it only makes sense that the resultant modulated waveform will also have two states—+1 and −1—which can directly set the state of the switch. If this was not the case, and the modulated waveform were a free analog signal, we would have to formulate another strategy to regulate the state of the switch, and this would simply add another layer in the entire process.

Now that we have placed this constraint that the resultant modulated waveform can have only two states (+1 and −1), we need to think of how a high frequency modulation process can generate this waveform. However, before we introduce such a process, let us think of how this modulated waveform will be. If it can have only two states, but must have a high frequency component, this can only imply that the resultant waveform will consist of high frequency pulses. And if the resultant modulated waveform had high frequency pulses, this will mean that the switch will change its state not just twice in a cycle, but many times—twice for every pulse. The switch is now being regulated at a high frequency compared to being regulated at the same frequency as the modulation signal in the previous section. With this inference made, let us now try to infer the process that will produce these high frequency pulses that constitute this modulated waveform.

In the previous section, we had produced the switching states by comparing the modulation signal with zero. If we were to produce high-frequency switching signals, we must compare the modulation signal not with zero, but with a high frequency waveform. In a manner similar to AM or FM, this high frequency carrier signal has a frequency that is much higher than the frequency of the modulation signal. However, unlike communications, the frequency of the carrier signal will affect the frequency at which the switch changes state, and therefore, has to be constrained by how fast the switch can operate. All power devices come with detailed specification sheets which describe the maximum frequency at which they can be switched, the minimum time they need to transition from cut-off to saturation and vice-versa. Therefore, though we would like the carrier signal to have a much higher frequency than the frequency of the modulation signal, we are limited by the physical capability of the switch.

Let us try and synthesize a waveform for the carrier signal. Usually for any particular application, one can think of more than one waveform. Due to the fact that the carrier signal has a much higher frequency than the modulation signal, during one cycle of the carrier signal, the modulation signal can be assumed to be constant. The next question is on the shape of the carrier signal. This again will change with the application, but to begin our discussion, if we wish to synthesize an ac waveform such as a sine waveform, we will need a carrier signal that will be symmetric in both positive and negative half-cycles of the waveform. In the basic example of AM and FM in the previous sections, the carrier signal

Fig. 2.12 Comparison between
carrier and modulation
signals—PWM

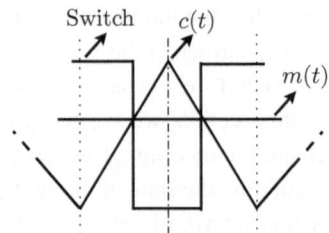

was a cosine waveform that was combined with the modulation signal. However, for power electronic applications, the carrier waveform will need to have linear segments that can be used to generate gating pulses. This is best described with an example. Let us consider the carrier signal to be a waveform with the same magnitude of unity as that of the modulation signal with the peak extending from −1 to +1 just like the modulation signal.

Figure 2.12 shows one cycle of a sample carrier waveform $c(t)$ [9–12]. This carrier waveform is a triangular waveform that has a magnitude of −1 at the beginning of the cycle. Its magnitude rises to +1 at the middle of the cycle and then drops back to −1 at the end of the cycle. As already stated, in this one cycle of the carrier waveform, the modulation signal $m(t)$ will appear as a constant since the frequency of the carrier waveform is much higher than the modulation signal. To achieve this switching, one can apply the following comparison:

When $m(t) > c(t)$; Switch = +1

When $m(t) \leq c(t)$; Switch = −1

Before we start describing the above process, let us make a few observations on the switching within the cycle of the carrier signal in Fig. 2.12. The switch has exactly two transitions in the cycle. Therefore, the switch undergoes one complete switching cycle as it changes its state within the cycle and reverts back to the original state before the end of the cycle. If one were to look at the switching with respect to the center line of the carrier cycle, the switching is symmetrical in the cycle across this center line. By choosing the carrier waveform to be a triangular waveform with a peak at the center of the cycle, we have managed to achieve this symmetry of switching. In the next section, we will examine another carrier waveform that does not meet this criterion. The higher is the value of the modulation index, the thinner will be region for which the switch is regulated to −1.

Let us implement this with a Python simulation as follows:

```python
import numpy as np
import matplotlib.pyplot as plt
from scipy import signal
from numpy.fft import fft, fftfreq

def compare(a, b):
    if a > b:
        return 1
    else:
```

```
            return −1
vcompare = np.vectorize(compare)

fm = 50.0
omega_m = 2∗np.pi∗fm
fc = 5000.0
omega_c = 2∗np.pi∗fc

t_duration = 1.0
t_step = 1.0e−6
fsampling = 1/t_step

time_array = np.arange(0, t_duration, t_step)
no_of_data = np.size(time_array)

# Modulation signal
mod_index = 0.8    # modulation index
mod_signal = mod_index∗np.cos(time_array∗omega_m)

# Carrier signal
carr_signal = signal.sawtooth(omega_c∗time_array, 0.5)
# The sawtooth waveform goes from (0, −1) to (pi, 1) for 0.5 duty cycle.

#PWM
switch_wave = vcompare(mod_signal, carr_signal)

# FFT of the square wave signal
freq_v_output = fft(switch_wave)/no_of_data
freq_array = fftfreq(no_of_data, t_step)

# Plot of the Sine PWM signal
plt.figure()
plt.plot(time_array, switch_wave)
plt.plot(time_array, mod_signal)
plt.plot(time_array, carr_signal)
plt.title('Result of Comparison')
plt.xlim([0.78, 0.785])
plt.xlabel('Time')

plt.figure()
plt.plot(time_array, switch_wave)
plt.plot(time_array, mod_signal)
plt.plot(time_array, carr_signal)
plt.title('Result of Comparison')
plt.xlim([0.783, 0.784])
plt.xlabel('Time')

# Plot of the frequency spectrum of the Sine PWM signal
plt.figure()
plt.plot(freq_array, np.abs(freq_v_output))
plt.title('Frequency response')
plt.xlim([0, 50000])
plt.xlabel('Hz')
```

```
plt.figure()
plt.plot(freq_array, np.abs(freq_v_output))
plt.title('Frequency response')
plt.xlim([0, 6000])
plt.xlabel('Hz')

plt.show()
```

The Python code for the example can be found in folder PWM within the folder chapter2 at the link: https://github.com/opensourceelectrical/switching-strategies-for-power-electronics.

Figures 2.13 and 2.14 show the simulation results. Figure 2.13 shows the modulation process over approximately quarter cycle of the modulation signal. It can be observed how the width of the switching pulses changes as the modulation signal changes. Figure 2.14 shows a close-up of the comparison process over a few cycles of the carrier signal. The

Fig. 2.13 Generation of switching status from PWM

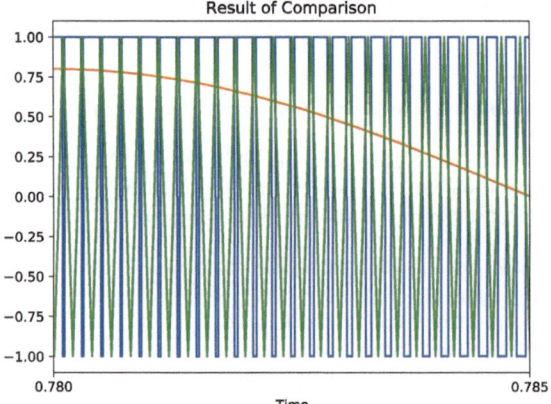

Fig. 2.14 Generation of switching status from PWM (zoomed)

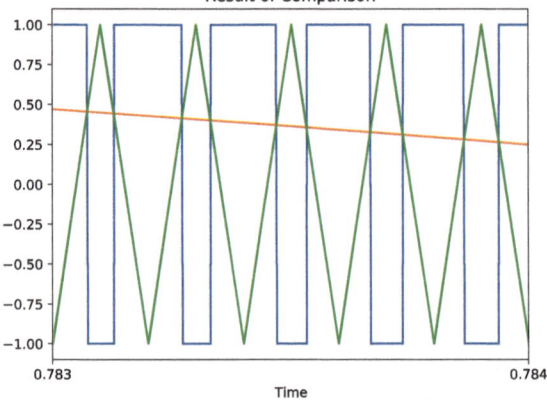

Fig. 2.15 Frequency spectrum of PWM waveform

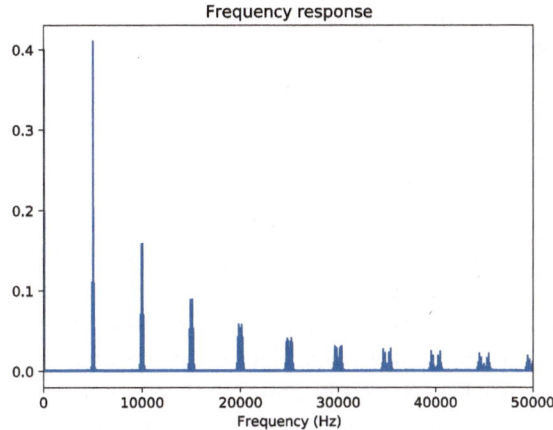

reader is encouraged to zoom into different parts of the waveform to examine how the pulse width changes over a cycle of the modulation signal. The reader can confirm that when the modulation signal is at its maximum positive peak of +1, the switch status is +1 for almost the entire carrier cycle, while for the minimum negative peak of −1, the switch status is −1 for the entire carrier cycle.

The reader can also verify through visual inspection that the width of switching pulses follow the low frequency sinusoidal envelope of the modulation signal even though the switching takes place at a high frequency.

Now that we have generated a modulated switching signal, let us perform a frequency analysis. Figures 2.15 and 2.16 show the frequency spectrum. Figure 2.15 shows the entire frequency spectrum, from which it can be observed that there are clusters of components at multiples of the carrier frequency of 5000 Hz. With increasing frequency, the magnitude

Fig. 2.16 Frequency spectrum of PWM waveform (zoomed)

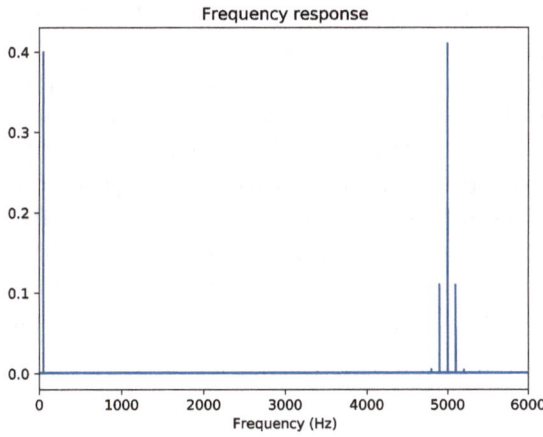

of the components decreases. Figure 2.16 shows a close-up of the frequency components until the carrier frequency. It can be observed that there is a prominent 50 Hz component, but no frequency components that are multiples of 50 Hz. At the carrier frequency of 5000 Hz, there are a number of side bands whose magnitudes decrease as they expand on either side of the carrier frequency component. These side bands are separated from the carrier frequency and from each other by the fundamental frequency of 50 Hz. On this point, there is a similarity with the frequency spectrum of the FM waveform.

To conclude on the result of pulse width modulation, let us sum up our observations. Due to the pulsed regulation of the switches, the resultant waveform has a 50 Hz component due to the low frequency envelope that is clearly evident, and has switching frequency components due to the high frequency pulses. However, unlike the previous section, there are none of the odd harmonics 150, 250, 350 Hz etc., nor are there any new harmonics such as the even harmonic or a dc component (0 Hz component). The resultant waveform also has very high frequency components, and these components have a higher magnitude as compared to the frequency spectrum of the previous section. Therefore, at first glance it appears that we have mixed results, and that modulation has solved some issues but has introduced others.

The biggest issue in the frequency spectrum of the previous section with rectangular switching was the presence of low frequency harmonics and the narrow separation between the lowest harmonic 150 Hz and the fundamental 50 Hz which is the frequency of interest. In this PWM simulation, this issue has been resolved as the lowest harmonic is a side band of the carrier frequency of 5000 Hz which has a far greater separation from the fundamental 50 Hz component. This separation makes filter design far simpler, less expensive and also less likely to fail due to ageing of components. Instead of designing a low pass filter with a cut-off frequency of around 80–100 Hz, we can now design a low pass filter with a cut-off frequency of several hundred Hz. Furthermore, a filter that effectively eradicated the frequency side bands around 5000 Hz, will be more effective for the frequency side bands present at multiples of the carrier frequency such as 10000, 15000 Hz etc. Therefore, the presence of higher frequency harmonics in the frequency spectrum does not pose an issue.

In this section, we have examined how modulation can be introduced in the context of power electronics to regulate the status of the switch. The effect of modulation is similar to the case of AM and FM, where the frequency spectrum can be seen to be transplanted to the high frequency regions around the carrier signal. In AM and FM, to extract the modulation signal at the receiving end, a demodulator is needed. In power electronics, one needs a filter to be able to extract the fundamental frequency corresponding to the modulation signal, and from the frequency analysis, it is very clear that PWM makes the process of designing a filter much simpler. In the next section, we will consider a very simple power electronic converter that will use a similar PWM strategy, and will show how filters and passive elements such as inductors and capacitors can result in shaping the final output waveform.

2.8   Step Down dc-dc Converter or Buck Converter

In the previous section, we had introduced the concept of PWM at a very abstract level using just the state of a switch which can attain two states +1 and −1. In this section, let us implement PWM in an actual power electronic converter and observe the resultant behaviour of the converter. A power electronic converter would use a power device such as MOSFET (Metal-Oxide-Semiconductor Field-Effect-Transistor) or an IGBT (Insulated Gate Bipolar Transistor), and furthermore, there will differences between manufacturers and also the material used (Silicon vs. Silicon-Carbide or Gallium Nitrate). However, in this book, we will consider ideal devices as the objective is to examine the effect of switching in general rather than delve into the details of device characteristics. Moreover, every power electronic converter will need a closed-loop control strategy to ensure that the output produced is regulated close to a desired reference value. However, in this book, we will consider every power electronic converter operating in open-loop without any output regulation, as control is a completely different domain which is out of the scope of this book.

Before we start with the actual converter, let us examine a few basics of ideal power devices. In this book, there will be usually two types of ideal devices—the uncontrollable (diodes) and the controllable (MOSFETS, IGBTs or other equivalent devices) [9–12]. A diode has two terminals—an anode and a cathode. Figure 2.17 shows the representation of the diode which will be found in circuit diagrams in this book. A diode will allow current to flow through it in only one direction—from the anode to the cathode as shown in the diagram. A diode will start conducting when it is forward biased—when the voltage appearing across the anode and cathode is positive and greater than a certain value called the junction potential which is in range of 0.3 V to a few Volts for high capacity devices. When the diode starts conducting, it will present a near constant voltage close to the junction potential called as the forward voltage drop. In this book, since we are considering ideal diodes, we will assume a diode to be forward biased when the voltage across the anode-cathode terminals is greater than zero, and when the diode starts conducting, it will be assumed to be a short-circuit.

Once a diode starts conducting, it will continue to conduct until the current exceeds a maximum value called the maximum forward current as this will destroy the device. Once the diode starts conducting, it will only stop conducting when the current through it reverses. In reality, a certain reverse current is needed to re-establish the barrier junction. However, for an ideal diode, a zero current or a minutely small negative current will be sufficient to stop its conduction. When a diode is reverse biased—when the anode-cathode terminal voltage is negative, the diode will conduct a reverse current, which is in the range of a few

Fig. 2.17 Diode

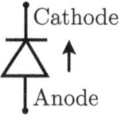

Fig. 2.18 Controllable power device

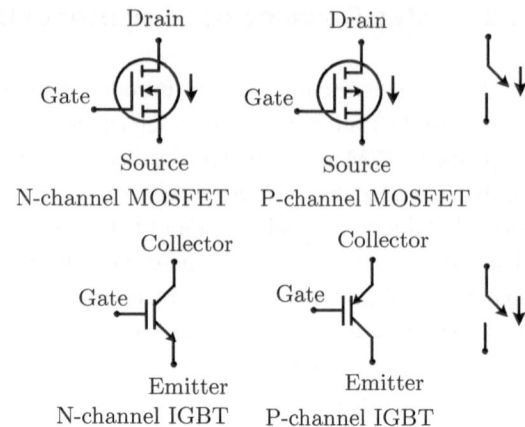

Drain — Gate — Source
N-channel MOSFET

Drain — Gate — Source
P-channel MOSFET

Collector — Gate — Emitter
N-channel IGBT

Collector — Gate — Emitter
P-channel IGBT

microamperes or less. For an ideal diode, it can be assumed that the diode will not conduct any current when reverse biased. If the reverse biased voltage is continuously increased beyond a certain value, the diode will be destroyed. Simulations will not examine the destruction of the diode as it is fairly complex, but when a diode is reverse biased, the voltage across it can be measured.

In the case of controllable devices, they have three terminals in contrast to the diode as shown in Fig. 2.18. There are generally two families of devices—one with emitter-collector-gate terminals (such as IGBTs) and the other with source-drain-gate terminals (such as MOSFETs). Additionally, one can have N-channel devices or P-channel devices depending on whether the majority carriers are electrons or holes. One can differentiate between them in the symbol with the arrow as shown in Fig. 2.18. However, the direction of current through them will not change. These devices differ vastly internally with respect to construction and can offer different characteristics. However, from an operational viewpoint, they are equivalent. Similar to diodes, these devices only conduct current in one direction through them—from collector to emitter or from drain to source. On the extreme right, is the ideal device representation which is similar to a switch, with the arrow on the switch indicating the direction in which the current can flow through the device.

For a device to conduct, it must be forward-biased i.e the voltage across the collector-emitter terminals or the drain-source terminals must be positive. Additionally, there must a positive voltage applied across either the gate-emitter terminals or the gate-source terminals. Without this voltage (called gate signal) across the gate-emitter or gate-source terminals, the device will not conduct even if it is forward biased. If the forward-bias voltage across the collector-emitter or drain-source exceeds a certain value called maximum forward blocking voltage in the absence of a gate signal, the device will be damaged. The device will conduct as long as the current through it flows in the allowed direction and a gate voltage exists across the gate-emitter or gate-source terminals. The device will stop conducting if the current through it reverses similar to a diode. The device will also stop conducting if a

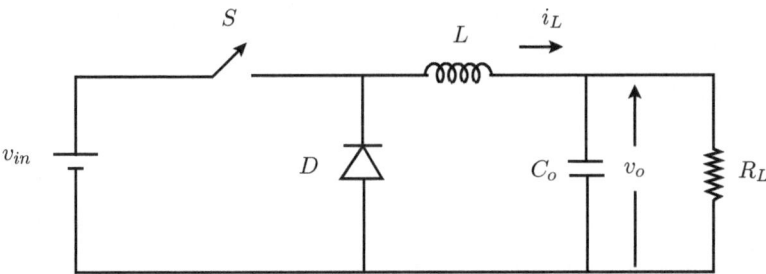

Fig. 2.19 Buck converter circuit

negative voltage is applied across the gate-emitter or gate-source terminals. In this aspect, these controllable devices differ from diodes as we can now turn them off at will. If however, the gate signal is removed while the device is conducting a current, the device will enter the active region and will be damaged. Therefore, they need a positive gate signal while in conduction mode to ensure that the device remains in saturation region.

With this background of power devices, let us consider the topology and operation of a step-down dc-dc converter, popularly called a buck converter [9–13]. Figure 2.19 shows the topology of a buck converter. In a practical circuit, there might be auxiliary circuits related to snubbers and protection, but, Fig. 2.19 shows the basic circuit from a conceptual standpoint. Such a buck converter is typically used when one needs a regulated output having a lower voltage level than the unregulated input. Let us consider an example when one has a 24 V battery (v_{in}), but would like to generate a steady 9 V dc supply (v_o), as the battery will discharge and then be recharged. As stated before, closed-loop control will not be covered in this book, and therefore, the simulation will merely choose an operating point that results in an output voltage close to 9 V. The output is formed across the capacitor C_o shown in the circuit, which in practice is usually an electrolytic capacitor, which forms a steady voltage across it even when the current flowing through it has significant ripple as will be shown soon. Across the output capacitor C_o, the load is connected depicted by R_L in the diagram, which in practice could be an electronic circuit, and R_L is merely a representation of the average power drawn by the load.

The rest of the components between the input voltage source and the output capacitor are what make the converter work. The power device such as a MOSFET is denoted by S. Before we begin with regulating the state of the power device S with PWM, let us examine the basic working of the converter. The converter has three states which can be best described as progressing from one state to the other. Let us begin with the converter completely dead—the output capacitor C_o has no charge and therefore output voltage is zero and no current is flowing in load resistor R_L. Also, let us assume that the inductor L is completely de-energized and the current through it is zero. If no gate signal is provided to the power device S, it will not conduct and no current will flow through it. The diode D will also not be forward biased and therefore will not conduct. This state where both the power device S and

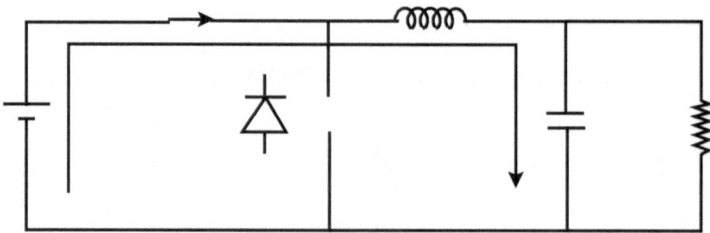

Fig. 2.20 Buck converter when device S is conducting

the diode D are not conducting is one possible state of the converter and in the following discussion, we will examine when it can occur, though here, it is the special case of a circuit "at rest".

If a gate pulse is provided to power device S, it will begin to conduct. This is because power device S is forward biased, as the input voltage v_{in} is approximately 24 V, while the output voltage is 0. Therefore with power device S forward biased, it will begin to conduct. This state is shown in Fig. 2.20. The KVL equation for the forward path can be written as:

$$v_{in} - L\frac{di}{dt} - v_o = 0 \tag{2.21}$$

Here, the output voltage v_o is 0, but by assigning it the variable, we have generalized the equation for when it can change later. Since v_o is initially 0, the current i in the above equation will increase.

Since a current is flowing through the output capacitor C_o, it will get charged and therefore the voltage across it will rise:

$$i - C_o\frac{dv_o}{dt} - \frac{v_o}{R_L} = 0 \tag{2.22}$$

In order for the output voltage v_o to rise, the only assumption to be made is that the load is not too high and therefore resistor R_L is not too low.

The diode D will be reverse biased because when power device S is conducting, the input voltage v_{in} appears directly across the diode but the anode-cathode voltage of the diode will be negative. Therefore, diode D will not conduct. This state of operation of the converter when power device S is conducting and diode D is not conducting is one principal state of the converter. This state is characterized by the current through the inductor increasing thereby storing energy in the inductor, and the voltage across the capacitor rising as it gets charged. It should be noted that the input voltage v_{in} is supplying energy to both the inductor as well as the output capacitor. Moreover, as the current through the inductor rises, the induced emf generated by the inductor will oppose the rise in current (according to Lenz's law) and therefore:

$$v_{in} - e_L - v_o = 0 \tag{2.23}$$

Due to which $v_o < v_{in}$, and the converter operates as a buck converter.

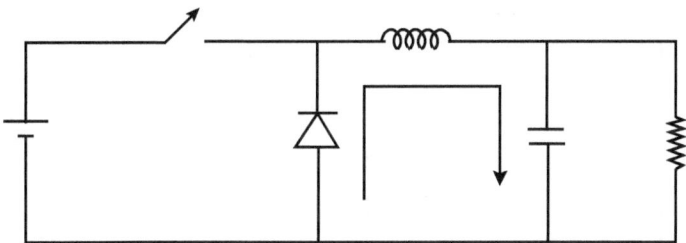

Fig. 2.21 Buck converter when diode D is conducting

When power device S is given a negative gate pulse and is turned off, it will stop conducting. As we know from basic electrical engineering, the current through an inductor cannot change instantaneously as the induced emf produced across the inductor is directly proportional to the change in the inductor current. Therefore, the induced emf will be such that it will attempt to force the inductor current to continue flowing even though the input voltage has been disconnected. This induced emf will force the diode to be forward biased and will start conducting to allow the inductor current to flow as shown in Fig. 2.21. This process is called freewheeling and is critical to the operation of many power converters. In this state, the equation for the current can be written as:

$$-L\frac{di}{dt} - v_o = 0 \tag{2.24}$$

Since the inductor which was receiving its energy from the input voltage is now no longer connected to a prime energy source, its current will start falling as is evident from the above equation. Since the inductor current is the prime supplier of energy to the output capacitor, the capacitor will also begin to discharge and the output voltage v_o will start falling. This state of operation of the converter, where power device S is not conducting but diode D is conducting, and where both the inductor current and output capacitor voltage are falling is the second principal state of operation of the converter. The question is how long can this state continue—either as long as power device S remains off or until the inductor current becomes zero.

If the inductor current becomes zero, the diode D will stop conducting. In this state, both power device S and diode D are not conducting, there is no inductor current and the output capacitor will continue to discharge as there is a load connected across it as shown in Fig. 2.22. This state of operation is a third possible state of the converter. The reason why it is possible is because it might not occur if the power device S is turned ON again by a positive gate pulse before the inductor current becomes zero. In such a case, once power device S starts conducting, the input voltage v_{in} appears directly in reverse across the diode, and will attempt to force a current in reverse across the diode. The diode will stop conducting, and with the power device S beginning to conduct, the inductor current will transfer to the power device S.

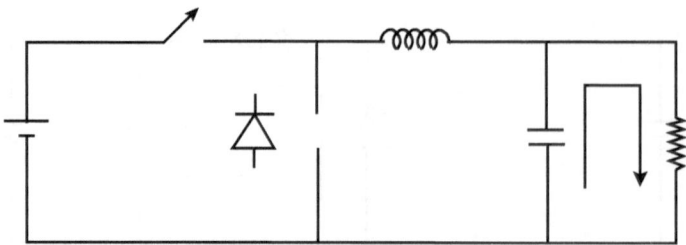

Fig. 2.22 Buck converter when both device S and diode D are not conducting

If we assume the case of power device S turning on before the inductor current becomes 0, we can apply the condition that in steady state, the increase in inductor current when power device S is conducting will be equal to the decrease when diode D is conducting. This is due to the fact that in steady state, the average power drawn being constant, the average current drawn by the load R_L will also be constant, and consequently, the average inductor current will be constant. During transients, such as when the converter is starting up or when the load is changing, this condition will not be true, as the average current through the inductor is changing. However, in steady state we can apply the condition:

$$t_{ON} \frac{v_{in} - v_o}{L} = t_{OFF} \frac{v_o}{L} \tag{2.25}$$

By multiplying the rate of change of current $\frac{di}{dt}$ in the two operating states with the interval of the two states—t_{ON} and t_{OFF} respectively, we have merely equated the change in current during the two intervals.

From the above:

$$\frac{v_o}{v_{in}} = \frac{t_{ON}}{t_{ON} + t_{OFF}} = \frac{t_{ON}}{T_0} \tag{2.26}$$

Which results in the definition of a term - duty ratio which is defined as:

$$d = \frac{t_{ON}}{t_{ON} + t_{OFF}} = \frac{t_{ON}}{T_0} \tag{2.27}$$

The duty ratio is merely the duty performed by the device S during a complete time interval T_0 that includes both the ON time of device S as well as the OFF time of device S.

A great deal of literature can be found on the operation and control of the buck converter [9–12]. The above description was to make our next implementation of PWM easier to understand. As we have seen above, the power device S must be turned on for a particular period and then turned off for another period. With the input voltage v_{in} being 24 V and the desired output voltage v_o being 9 V, the duty ratio above needs to be:

$$d = \frac{9}{24} \approx 0.38 \tag{2.28}$$

It should be noted that for the power converter to operate, the power device S will have to be turned on and off in cycles, as when it is turned on, the output voltage will increase, and when turned off, the output voltage will decrease. Therefore, the output voltage will never be exactly 9 V, but will have a ripple around 9 V.

The question then arises—how do we wish to perform this turning on and off? Similar to the previous section where we regulated the switch as a rectangular waveform to have the same time period as a sine waveform, we can perform it at low frequency, since after all we need only a dc voltage (0 Hz). However, a low frequency of operation will imply that the output capacitor will have to be large to prevent the ripple from being too high. This is due to the fact that when device S is on, the capacitor will charge, but to prevent it from charging to too high a voltage, the only solution is to increase the value of the capacitor. Mathematically, this is similar to our frequency analysis before. We are interested in a dc (0 Hz) output, but, besides this output, there will be other frequencies that we need to deal with. Operating the device S at low frequency will result in low frequency components in the output, and these will need a large inductor and a large capacitor to filter them out.

PWM can be used for such a dc-dc converter as well. The modulation signal will be the duty ratio $d = 0.38$, as that follows from the principle of operation of the converter. We need to choose a carrier frequency that is of significantly high frequency, which for the sake of an example we can choose as 10000 Hz or 10 kHz. We need to devise a strategy of comparison with the modulation signal, so as to turn ON the device S at a particular instant and turn it OFF at another instant. There is a difference with the strategy shown in the previous section. Since in the previous section, we were generating an ac output, we needed a carrier that was symmetric in each cycle, such that the comparison would lead to a symmetric switch status. In the case of a dc-dc converter, we do not need such symmetry, as we are generating a dc voltage anyway.

The comparison between the carrier and the modulation signal can be shown in Fig. 2.23. The carrier waveform shown in Fig. 2.23 is called as a sawtooth waveform. The reader is encouraged to compare this sawtooth waveform with the triangular waveform in the previous section. The sawtooth waveform allows a convenient translation between the duty ratio d and the time interval for which the device will remain turned ON due to the linear rise from 0 to 1 over the switching cycle. From Fig. 2.23 is is clear that the device S will be turned ON at the beginning of every carrier cycle. The device S will be turned OFF when the carrier

Fig. 2.23 Comparison between carrier and duty ratio

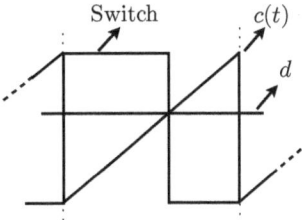

signal is greater than the modulation signal, and will remain turned OFF until the end of the
cycle. This strategy can be implemented with the following Python code block.

```python
sw_freq = 10000.0
# Control time interval
dt = 1.0e-6
if t_clock > t1:
    carr_signal += (1 / (1 / sw_freq)) * dt
    if carr_signal > 1.0:
        carr_signal = 0.0
    mod_signal = 0.38
    if mod_signal > carr_signal:
        s1_gate = 1.0
    else:
        s1_gate = 0.0
    t1 += dt
```

In the above code block, the variable `t_clock` is the time instant of simulation provided
automatically by the circuit simulator in any control file, `carr_signal` is a static variable
whose memory is stored between iterations and `s1_gate` is an output variable which
directly interfaces to a controllable component representing the power device. The reader is
encouraged to read the resources on the circuit simulator in the introduction of the book for
more details on these terms.

We can simulate a buck converter using Python Power Electronics. The simulation can
be found in folder `buck_converter` within the folder `chapter2` at the link: https://
github.com/opensourceelectrical/switching-strategies-for-power-electronics.

Figures 2.24, 2.25, 2.26 and 2.27 show the output voltage, the current through the inductor
and the ripple in the inductor current and the output voltage. Figure 2.28 shows the PWM
process.

The reader is encouraged to plot the other variables to understand the operation of the
converter. However, since the objective of the book is not to describe in detail the behaviour

Fig. 2.24 Buck converter
output voltage

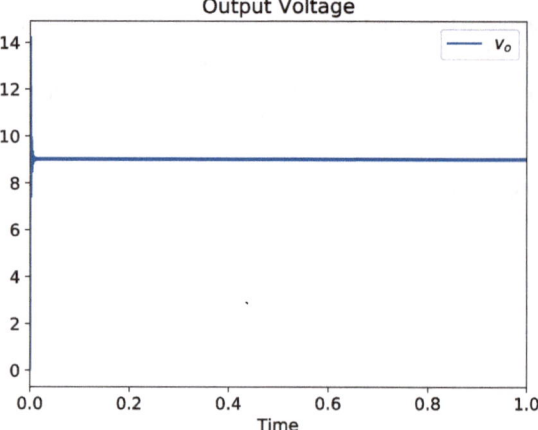

Fig. 2.25 Buck converter output voltage (zoomed)

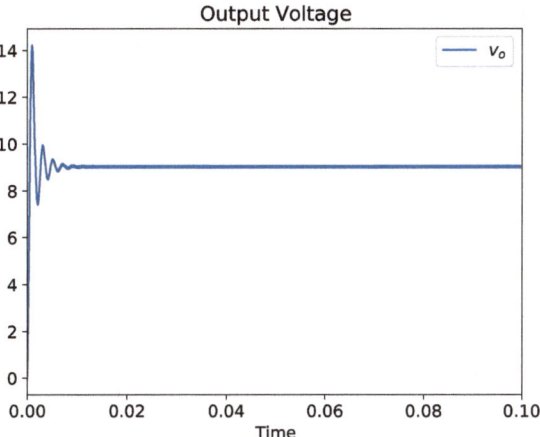

Fig. 2.26 Buck converter inductor current

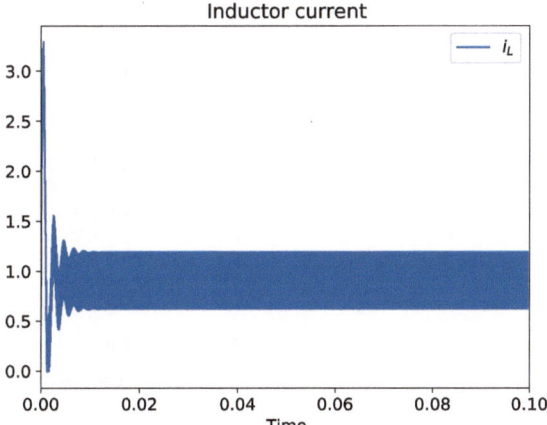

of the buck converter, but rather to understand the effect of switching, we shall progress towards frequency analysis of the waveforms. Since the two most crucial waveforms in our analysis are the switched voltage produced immediately after the power device S and the final output voltage v_o, we shall use these two variables in our frequency analysis programs from the previous sections.

To perform frequency analysis, one can use two approaches—to perform them in real-time as a part of the simulation, or to perform them offline after the simulation has completed. Most simulators allow users to perform frequency analysis while the simulation is in progress. However, unless this frequency analysis will be actively used in the real-time control of the system, it is an unnecessary burden to continuously perform frequency analysis while the simulation is running. In this book, we will perform frequency analysis on simulation data after it has completed. In order to do so, a few steps need to be performed to extract the data into Numpy arrays so that the code from the previous sections can be used.

Fig. 2.27 Buck converter
inductor current and output
voltage ripple

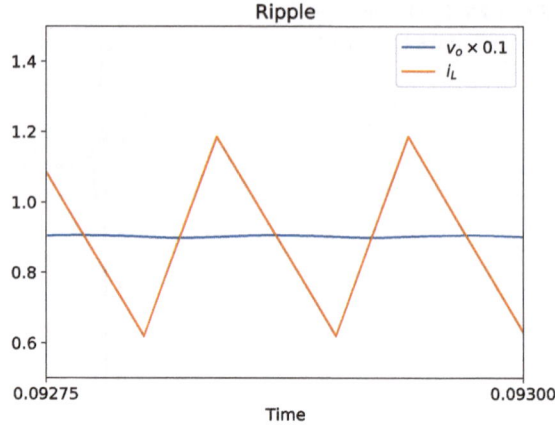

Fig. 2.28 Buck converter
PWM

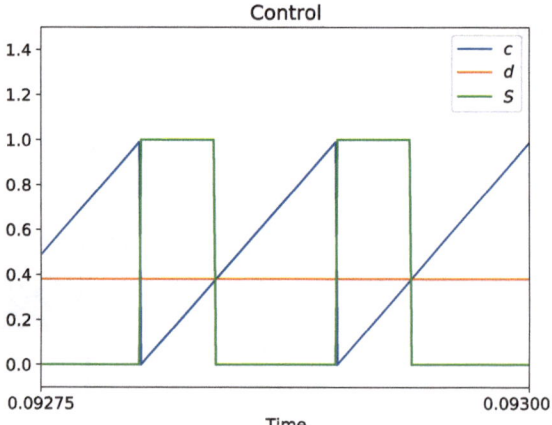

The simulation data is written into a text file which is specified in the simulation parameters and this output text file exists in the same working directory as the circuit schematic files. To extract the data into Numpy arrays, we use the following code:

```
import numpy as np
import matplotlib.pyplot as plt
from numpy.fft import fft, fftfreq
import pandas as pd

df = pd.read_csv(
        ''ckt_output.dat",
        delim_whitespace=True
    )
df.columns = ['Time', 'i_in', 'i_L', 'i_d1', \
            'i_load', 'i_cout', 'vsw', 'vo', \
            'm', 'c', 'g']

time_array = np.array(df['Time'])
vo = np.array(df['vo'])
```

```
vsw = np.array(df['vsw'])

no_of_data = len(time_array)

freq_v_sw = fft(vsw)/no_of_data
freq_array = fftfreq(no_of_data, time_array [1] − time_array[0])
freq_v_output = fft(vo)/no_of_data

plt.figure()
plt.plot(time_array, vo)

plt.figure()
plt.plot(freq_array, np.abs(freq_v_sw))
plt.title('Frequency response')
plt.xlabel('Hz')

plt.figure()
plt.plot(freq_array, np.abs(freq_v_output))
plt.title('Frequency response')
plt.xlabel('Hz')

plt.show()
```

Using Pandas, one can read the file using the `read_csv()` method. Our output file is not in the .csv format (Comma Separated Value), but rather is a mere text file. However, we still write data to it in a column format with each column separated from the other using mere spaces and not commas. For this reason, it is necessary to provide the argument `delim_whitespace=True` as otherwise the `read_csv()` method will look for commas. The `read_csv()` method will read the contents of the output data file and populate it in a Pandas dataframe which one could say is a table with certain advanced functionalities. This table will be provided with default column names. However, since the simulator does not write column names into the output data file, the column names will be randomly chosen by Pandas according to the values in the columns. Our first task is to rename the columns according to the actual data being written to the output data file by the simulator.

To determine which columns correspond to which meter outputs and control variables, the simulator produces a key file called plotkey.txt when the simulation is run. In this file, one can find which variable corresponds to which column. As an example, for this particular simulation, the contents of this file are:

```
t = 1
Ammeter_in = 2
Ammeter_L1 = 3
Ammeter_D1 = 4
Ammeter_load = 5
Ammeter_Cout = 6
Voltmeter_switching = 7
Voltmeter_Vo = 8
pwm_mod_signal = 9
pwm_carr_signal = 10
gate_signal = 11
```

Every variable, meter and control variable that appears in the output data file has been assigned a column number. We use this to rename the columns. It is important to note that we must specify names for each column in the dataframe when renaming the columns. Once this is done, we can access any column by specifying the column name with the dataframe variable name. We can also convert the column data into a Numpy array using `np.array()` method. In the previous sections, we had defined the number of data points. In this case, we must determine the number of data points, which can be done using the Python `len()` method. For the frequency range, we need the time interval between the data points. In the previous section, we had set this manually as 1 microseconds when generating data points. In this simulation, we know that the simulation integration time step as well as the interval of data storage is 1 microsecond. However, it is better not to assume this as one can always change the simulation parameters at will. Therefore, the time interval is calculated as the difference between any two samples of the `time_array`—which can be the 0th and 1st sample.

Following these computations, the remaining frequency analysis is the same as the previous sections. We will perform frequency analysis on the switched voltage following the power device S and on the output voltage v_o. Figures 2.29, 2.30 and 2.31 show the results. As can be seen, the switched voltage has a large range of frequency components. We are interested in the dc component (0 Hz) while other components will need to be removed. As can be seen, the other frequency components are harmonics of the switching frequency 10 kHz. The frequency spectrum of the output voltage shows only the dc component to be remaining and all the switching frequency components are effectively removed. This is evident from the nature of the output voltage waveform in Fig. 2.25. Therefore, the inductor L and the capacitor C_o have formed a very effective low pass filter to remove the nuisance harmonic components from the final output voltage. The reader is encouraged to repeat this simulation while decreasing the switching frequency of the converter and examine the results.

Fig. 2.29 Frequency analysis of the switched voltage

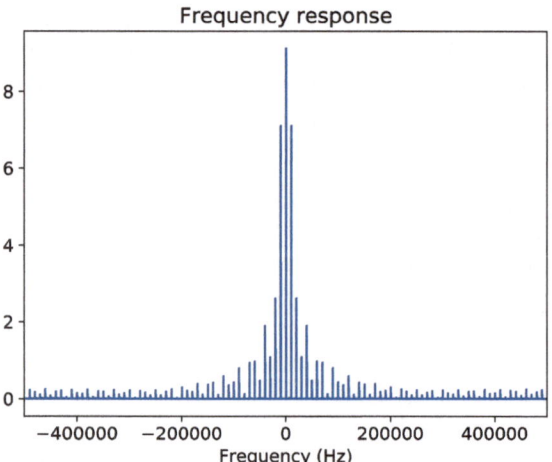

Fig. 2.30 Frequency analysis of the switched voltage (zoomed)

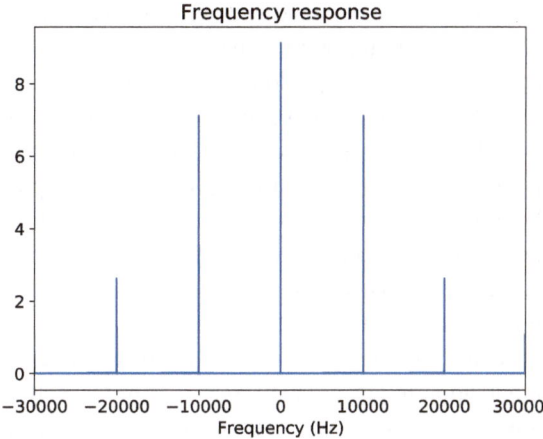

Fig. 2.31 Frequency analysis of the output voltage

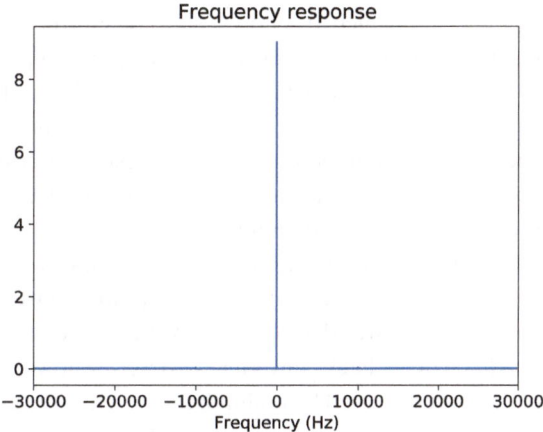

2.9 Conclusions

In this chapter, we have examined the concept of modulation in detail, and with examples from outside power electronics. The reason behind this non-power discussion is due to the fact that signal processing quite often precedes power electronics in the undergraduate curriculum, and with communication devices such as mobile phones and WiFi now ubiquitous, an understanding of modulation in communications is a bit more natural than in power electronics. With the discussion of amplitude modulation and frequency modulation, we describe how it is possible to mix a low frequency signal and a high frequency signal to result in a modulated waveform that has a new frequency spectrum. This is fascinating to some extent similar to how in chemistry, one can mix two very different elements to produce a molecule with drastically different properties.

In order to understand and verify the concept of frequency transformation due to modulation, we started with some very basic trigonometric waveforms, and verified that the resultant frequency spectrum matches the theoretical result. We introduced very convenient in-built methods available with Numpy to perform frequency analysis. To fully understand how these methods worked, we introduced frequency analysis through Fourier Series and Fourier Transform. We also introduced the digital equivalents of the Fourier Series and Fourier Transform, and compared them with the result of the Numpy methods and interpreted the frequency spectrum plots. Though these are dealt with in detail in basic signal processing courses, there is still a gap in translating these concepts to power electronics, and this chapter fills this gap.

With a background of modulation, we have then gradually introduced the concept of modulation in power electronics. We began with a very basic abstract level description of how a simple switch can be used to synthesize a sine waveform. This discussion contained no power electronics, but merely examined the frequency spectrum of a switch regulated in a logical manner to follow a sine waveform. The presence of low frequency harmonics in the frequency spectrum of a rectangular approximation of a sine waveform gave an indication of the burden in filtering out the nuisance frequency components and producing the desired sine waveform. Following this description, we described Pulse Width Modulation, and using a simulation and code samples, described how this can be implemented. The resultant frequency spectrum shows a similar frequency pattern as compared to Frequency Modulation used in communications.

We finally presented the case of a step-down dc-dc converter or buck converter. The buck converter is one of the simplest power converters to understand as it contains a single controllable power device and only a few operating states. After a basic description of the operation of the buck converter, we describe how PWM can be implemented in the switching of the power device. Using simulation results, we describe how the output voltage is relatively smooth. We adapt our frequency analysis present in the previous examples to be applicable to the results of a simulation, and find that the frequency spectrum of the switched voltage waveform contains a wide range of switching frequency harmonics, but, the output voltage waveform retains only the dc (0 Hz) component.

References

1. A.V. Oppenheim, A.S. Willsky, S.H. Nawab, *Signals and Systems* (Prentice Hall, 1997)
2. M. Schwartz, W. Bennett, S. Stein, *Communication Systems and Techniques* (Wiley, New York, 1995)
3. S. Haykin, *Communication Systems* (Wiley India Pvt. Limited, 2008)
4. B.P. Lathi, *Modern Digital and Analog Communication Systems* (Oxford University Press, Inc., 1995)
5. E. Brigham, *The Fast Fourier Transform and Its Applications* (Prentice Hall)
6. J. Unpingco, *Python for Signal Processing* (Springer International, 2014)
7. A.B. Downey, *Think DSP—Digital Signal Processing in Python* (Green Tea Press, 2014)

8. A. Zúñiga-López, C. Avilés-Cruz, Digital signal processing course on Jupyter-Python notebook for electronics undergraduates. Comput. Appl. Eng. Educ. **28**(5), 1045–1057 (2020) [Online]. Available: https://onlinelibrary.wiley.com/doi/abs/10.1002/cae.22277

9. N. Mohan, T.M. Undeland, W.P. Robbins, *Power Electronics: Converters, Applications, and Design*, 3rd edn. (Wiley, New York, 2002)

10. R.W. Erickson, D. Maksimovic, *Fundamentals of Power Electronics* (Springer Science & Business Media, 2007)

11. M.H. Rashid, *Power Electronics: Circuits, Devices, and Applications* (Pearson Education India, 2009)

12. D.G. Holmes, T.A. Lipo, *Pulse Width Modulation for Power Converters: Principles and Practice*, vol. 18 (Wiley, New York, 2003)

13. S. Roberts, *DC/DC Book of Knowledge* (RECOM Engineering GmbH, 2016)

Coordinated Modulation Strategies—Converters with Two Controllable Power Devices

3

3.1 Introduction

In the previous chapter, we had introduced the concept of Pulse Width Modulation (PWM) after presenting a background of modulation in the context of signal processing and communications. We had examined how PWM can be implemented in a basic power electronic converter such as a buck converter. The buck converter contains a single controllable device which can be switched using PWM, and therefore, is an extremely simple converter. In practice, power electronic converters can contain several controllable devices, and these devices will need to be switched in a coordinated manner. In this chapter, we will begin examining a few basic converters which have two controllable devices.

In a power converter with a single controllable device, we have only one degree of freedom—we can either turn ON or turn OFF the device. In the buck converter that we examined in the previous chapter, we found three possible states of conduction—the first where the controllable device was conducting, the second where the freewheel diode was conducting and the third where neither the controllable device nor the diode were conducting. The third state of no conduction was an unintended state that occurred due to the fact that the inductor current had a small magnitude and reached zero before the end of the switching cycle. However, the main two states were the first two states. This follows from our knowledge of binary digits (bits)—a single bit allows two states, namely 0 and 1. If we increase the number of bits, the possible combinations are expressed as 2^N, where N is the number of bits. Therefore, for the specific case of two bits, we would have $2^2 = 4$ combinations.

In a power converter with two controllable devices, we can therefore expect a total of four major possible states. Once again, we stress on the term "major" as it is possible to have some unintended states. If we list these states, we have a state where both controllable devices are conducting, two states where one of the devices is conducting, and a last state where none of the controllable devices are conducting. In the case of the buck converter,

S. V. Iyer and M. N. Aalam, *Switching Strategies for Power Electronic Converters*,
Synthesis Lectures on Power Electronics, https://doi.org/10.1007/978-3-031-41405-3_3

we had used a single freewheel diode in addition to the controllable device. In a converter with two controllable devices, the number of diodes would need to be determined by the exact topology of the converter, as the diodes are needed to ensure that the current through inductors is not abruptly broken. Therefore, if we take into consideration that we have more than one diode, we now have introduced more potential conduction states, as the controllable devices may conduct on their own or in combination with one or more freewheel diodes. From this discussion, it becomes clear, without delving into the details of converter circuits, if we use simple mathematics to determine the number of conduction states that might be possible, having even two controllable devices can result in a number of possible conduction states.

In this chapter, we will present a number of power converters that can be realized using two controllable devices. We will begin by presenting a buck-boost converter that is a modification of the buck converter described in the previous chapter. Though, there exist a number of ways to arrive at a buck-boost converter, the objective of this converter is to show how one can achieve more functionality with one more controllable device and one more freewheel diode. The operation of this converter is described along with a simulation. The reader can clearly examine how from all the mathematically possible conduction states, only a few occur. This is due to the fact that not all conduction states are possible, as at times when a controllable device conducts, other devices and diodes may not be able to conduct, or for a particular controllable device to conduct, another controllable device or diode must also simultaneously conduct.

This chapter will then introduce the concept of a converter leg which consists of multiple controllable devices and freewheel diodes. A number of these converter legs are available from several manufacturers, and they offer the advantage of a compact module containing multiple controllable devices and diodes connected in different topologies. The chapter, however, begins with the simplest and most popular module consisting of two controllable devices connected in series, and each controllable device having a diode connected in parallel to it, but in such as manner that the diode will conduct in a direction opposite to the controllable device. These diodes are called anti-parallel diodes and permit the flow of current in both directions through the parallel combination of controllable device and diode.

The chapter will present many topologies that can be realized using a single converter leg. In many cases, the use of these converter legs enables bidirectional flow of power through the converter. The chapter will show how a buck and buck-boost converter can be realized using converter legs. To describe how these converter legs can be used as building blocks for various applications, the last section will present a half-bridge dc-ac converter, where an ac system is supplied from a dc voltage source using a single converter leg. While presenting a converter topology, the operation of the converter will be described using circuit diagrams as well as simulation results. The reader should be familiar with simulating power electronic circuits using Python Power Electronics, as well as writing control files using Python in order to take full advantage of the contents of this chapter.

3.2   Modified Buck-Boost Converter

In this section, we will get started with a fairly simple topology for a buck-boost converter [1] in which the output voltage can be higher or lower than the input voltage. A buck-boost converter can be arrived at through many topologies, and a reader can find plenty of literature on this topic [2, 3]. The topology presented in this section is merely an extension of the buck converter topology already described and simulated in the previous chapter, with the boost stage being a minor modification to the buck converter. The objective behind presenting this topology is to describe how the controllable power devices will need to be coordinated in their conduction cycles for the converter to operate correctly. The purpose behind this section is also to describe how power converters can be synthesized in various different ways for the same end-result while having different advantages and disadvantages.

One simple realization of a buck-boost converter can be found in many power electronics textbooks using just a single controllable power device [2, 3]. However, that topology has the drawback that the output voltage has a reverse polarity as compared to the input voltage, due to which, the converter cannot have a single common ground terminal. By modifying the buck converter of the previous chapter, we can arrive at a buck-boost converter [1] whose output voltage has the same polarity as that of the input. The disadvantage is that this modified topology has two power devices instead of just one. Besides this one obvious difference, each topology requires a certain type of switching strategy which can result in different ripples in inductor currents and capacitor voltages. This in turn, will need them to be sized appropriately, which will affect the cost of the converter. In designing a converter, a power electronics engineer needs to take into consideration all these factors when comparing topologies - number of devices, rating of components, losses in components and potential of failure due to adverse stresses.

Figure 3.1 shows the topology of the modified buck-boost converter being simulated in this section [1]. There are two controllable power devices (MOSFETs or IGBTs) and they have been named as S_1 and S_2, while there are two diodes D_1 and D_2. The converter can be operated purely as a buck converter, and the same three operating states of the previous section can also be repeated here. In this case, the power device S_2 is never turned ON and

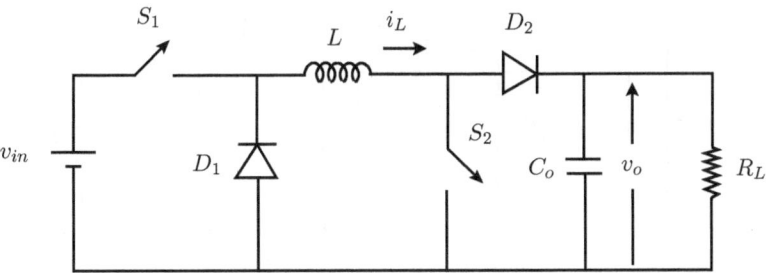

Fig. 3.1 Buck-boost converter circuit

therefore, never conducts, while device S_1 can be regulated according to a duty ratio. The diode D_2 will always be conducting when the converter operates as a buck converter. The reader can read the operation of the buck converter in the previous chapter. For the converter to operate as a boost converter, where the output voltage is greater than the input voltage, the power device S_2 will need to be operated in addition to the power device S_1. Since the power device S_1 brings the input voltage v_{in} into the circuit, this device will have to be turned ON completely during a switching cycle. During the switching cycle, the power device S_2 will be turned ON for a particular period of time and turned OFF for the rest of the switching time period similar to how the device S_1 was operated for buck converter operation. Therefore, the co-ordination between the devices is clear—for buck operation S_1 is switched while S_2 remains completely OFF, and for boost operation S_2 is switched while S_1 is completely ON.

Figure 3.2 shows the first operating state of the converter in boost mode. When the power device S_2 is turned ON (while S_1 is conducting), the inductor current increases according to the expression:

$$v_{in} - L\frac{di}{dt} = 0 \tag{3.1}$$

With the inductor current increasing, the energy stored in the inductor is increasing. The induced emf produced by the inductor will oppose this increase in current according to Lenz's Law:

$$v_{in} - e_L = 0 \tag{3.2}$$

Since the power device S_2 is conducting, the diodes D_1 and D_2 become reverse biased. The input voltage v_{in} appears across diode D_1 which reverse biases the diode, while diode D_2 is effectively connected in parallel across the output capacitor but the voltage across the anode and cathode of the diode is negative. The output capacitor C_o therefore receives no energy from either the input voltage or from the inductor, and therefore, the capacitor discharges through the load resistor, and the output voltage v_o decreases.

$$-C_o\frac{dv_o}{dt} - \frac{v_o}{R_L} = 0 \tag{3.3}$$

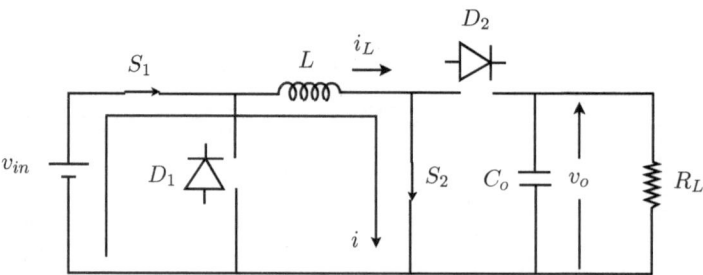

Fig. 3.2 Buck-boost converter when devices S_1 and S_2 are conducting

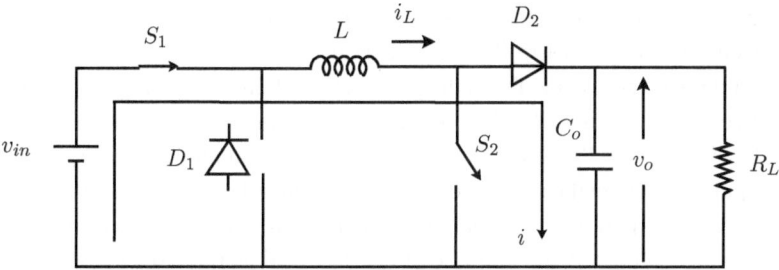

Fig. 3.3 Buck-boost converter when device S_1 and diode D_2 are conducting

Figure 3.3 shows the second operating state of the converter. When the power device S_2 is turned OFF (but S_1 continues to conduct), the energy in the inductor will now need to find a way to keep the inductor current flowing. The inductor current will once again freewheel through the diode. However, in contrast to the buck converter operation, the freewheeling will take place with the input voltage in the loop.

$$v_{in} - L\frac{di}{dt} - v_o = 0 \tag{3.4}$$

If one compares the conduction states of Figs. 3.2 and 3.3, it might appear that the inductor current will continue to increase even after the device S_2 is turned OFF, as the inductor is still fed by the input voltage source v_{in}. To fully understand the operation of the converter, one needs to separate the transient operation with steady-state operation. During the initial transient, when the output voltage v_o across the output capacitor C_o is increasing, it is clear from the current equations, that the rate of rise of inductor current will be higher when the device S_2 is conducting when the $\frac{di}{dt}$ is equal to $\frac{v_{in}}{L}$, as opposed to when the device S_2 is not conducting when the $\frac{di}{dt}$ is equal to $\frac{v_{in}-v_o}{L}$. As per Lenz's Law, the induced emf produced by the inductor will be such that it opposes the change in the current. Therefore, even though the current continues to rise during the initial transient, due to the decrease in the $\frac{di}{dt}$ when the device S_2 is not conducting, the induced emf produced across the inductor when the device S_2 is not conducting will be such that it attempts to force the inductor current to continue increasing at the higher $\frac{di}{dt}$ when the device S_2 was conducting. Therefore, considering the induced emf, one can write the expression:

$$v_{in} + e_L - v_o = 0 \tag{3.5}$$

As the output voltage v_o increases, the effect of the induced emf becomes more pronounced and causes the output voltage v_o to exceed the input voltage v_{in}. Under this circumstance, the $\frac{di}{dt}$ when the device S_2 is not conducting becomes negative and the inductor current begins to fall. However, the above equation still holds true, as according to Lenz's law, the induced emf produced across the inductor will be such that it opposes the decrease in inductor current in contrast to the time interval when device S_2 is conducting, and the inductor current was increasing. From the perspective of understanding energy flows, in the buck-boost converter of Fig. 3.1, when device S_2 conducts, the inductor is energized by the input, while when device S_2 stops conducting, both the input and the energy in the inductor feed the output resulting in the output voltage being higher than the input voltage. The reader should compare this operation with that of the buck converter in the previous chapter, where the input provided energy to the output capacitor as well as the inductor, due to which the output voltage was lower compared to the input voltage.

The capacitor will also charge as it receives energy from the input and the inductor:

$$i - C_o \frac{dv_0}{dt} - \frac{v_o}{R_L} = 0 \tag{3.6}$$

Even though the inductor current is decreasing, the current still supplies the output capacitor, whereas in the period with S_2 conducting, the output capacitor was disconnected from the input and the inductor. As with the case of the buck converter in the previous chapter, one must ensure that the load resistor R_L is not too small, as in that case, the output capacitor will discharge faster than it charges.

As with the case of the buck converter, it is possible that the inductor current will fall until it reaches zero. At that stage, the diode will stop conducting, and it is the capacitor alone that supplies the load causing the output voltage to start falling again. Whether this operating state occurs depends on whether the switch S_2 is turned ON again before the inductor current falls to zero. If we assume that the inductor current never falls to zero, we can in steady state, equalize the ripple in the inductor current over the two time intervals - when the switch S_2 is conducting and when it is not conducting:

$$\frac{v_{in}}{L} t_{ON} = -\frac{v_{in} - v_o}{L} t_{OFF}$$
$$v_{in}(t_{ON} + t_{OFF}) = v_o \, t_{OFF}$$
$$\frac{v_o}{v_{in}} = \frac{t_{ON} + t_{OFF}}{t_{OFF}} = \frac{T_0}{T_0 - t_{ON}} \tag{3.7}$$

With the duty ratio d being the ratio of the time interval for which the device S_2 is conducting in an entire switching cycle:

$$d = \frac{t_{ON}}{t_{ON} + t_{OFF}} = \frac{t_{ON}}{T_0} \tag{3.8}$$

Therefore, the final expression for the output voltage can be written as:

$$\frac{v_o}{v_{in}} = \frac{1}{1 - d} \tag{3.9}$$

Interestingly, for boost operation, the expression for the output voltage is almost the reciprocal of the expression for the output voltage in buck operation. Moreover, one needs to have completely separate definitions of the duty ratio for the buck and boost operations. In buck mode, device S_1 alone operates while device S_2 does not conduct, and in boost mode, device S_2 alone operates while device S_1 continuously conducts. This shows how versatile a power converter can be in terms of the output it can produce. In both modes, a few basic concepts need to be adhered to. The current through the inductor must have a path to flow at all times—whether a power device is conducting or not conducting. Either the inductor receives energy from the input and the inductor current increases, or the inductor is discharging its energy as the current freewheels.

In the case of the buck converter in the previous chapter, and also in the case of the buck-boost converter above, we have equated the ripple in the inductor current with respect to the time intervals for which the power device conducts and when it does not. This is the case during steady state, when the average of the inductor current is equal to the average current drawn by the load resistor R_L. For the average to remain constant, the increase in the inductor current when the power device conducts has to be equal to the decrease in the inductor current when the power device does not conduct. The ripple current will flow into the output capacitor that acts as a filter. Therefore, the equality that produces the expression for the output voltage with respect to the input voltage and the duty ratio, is merely the result of applying the rule of conservation of energy.

The simulation files for the buck-boost converter can be found in folder `buck_boost_converter` within the folder `chapter3` at the link: https://github.com/open sourceelectrical/switching-strategies-for-power-electronics.

The control file `gate_signal.py` contains control code for both buck mode as well as boost mode. Since, no closed loop control is implemented, these two modes are separated, though, in the presence of a closed-loop scheme, the mode of operation will be chosen through control rather than randomly assigned. Figures 3.4, 3.5 and 3.6 show the simulation results of the buck-boost converter. Figure 3.4 shows the output voltage which is clearly greater than the input voltage v_{in} which is 24 V. The inductor current shown in Fig. 3.5 has an average value in steady state that is equal to the average current drawn by the load resistor. The ripple in the output voltage and the inductor current can be observed in Fig. 3.6. The reader is encouraged to zoom into these results and verify that when the inductor current rises, the output voltage decreases. This is in contrast to the operation of the buck converter in the previous chapter, but is in line with the description of the operation of the converter above.

This section described a topology of a buck-boost converter using two controllable devices and two diodes. The readers who have a background of basic power electronics will be familiar with another commonly studied topology of the buck-boost converter that requires only one controllable device and one diode. The reader is encouraged to simulate the conventional buck-boost converter and the converter presented in this section and compare their performance. In this manner, one can think of various modifications to any converter that

Fig. 3.4 Output voltage of buck-boost converter in boost mode of operation

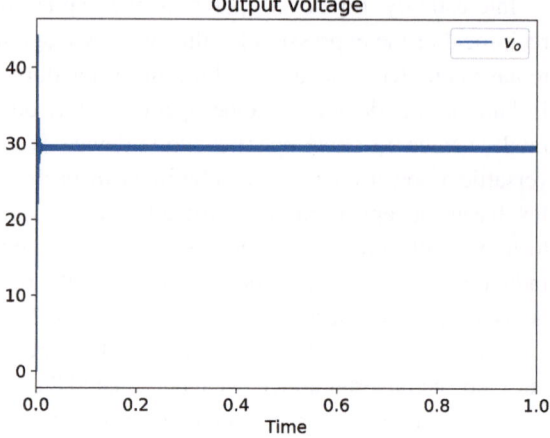

Fig. 3.5 Inductor current of buck-boost converter in boost mode of operation

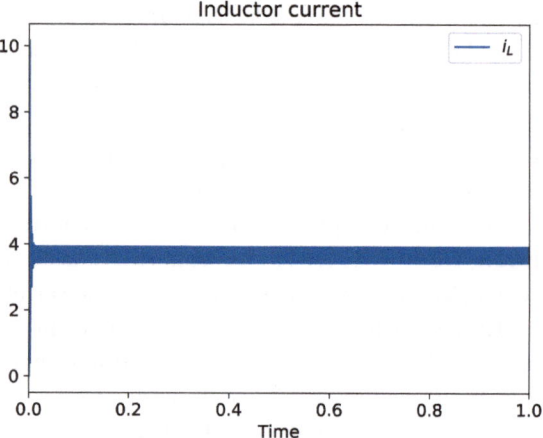

Fig. 3.6 Ripple in inductor current and output voltage of buck-boost converter in boost mode of operation

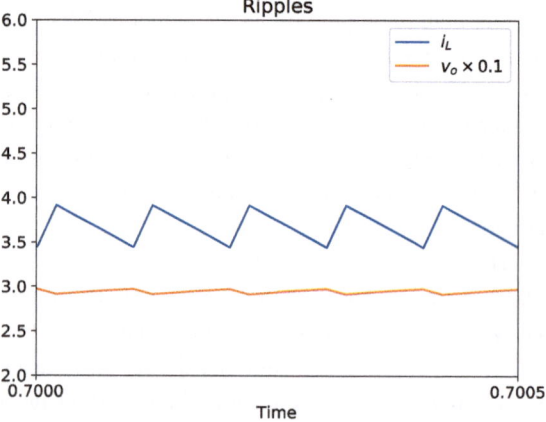

can introduce improvements or new features. These changes can be in the form of additional devices as well as additional inductors or capacitors. The reader is encouraged to perform a literature survey of the research publications in the domain of dc-dc converters and examine many of the novel topologies presented by researchers. In many of these novel topologies, any change in the topology usually requires a modification to the manner in which the devices are controlled, as is very clear in the case of the buck-boost converter presented in this section. The next section will examine a commonly used structure with multiple controllable devices and diodes which brings a great deal of convenience along with constraints on how the devices can be allowed to conduct.

3.3   A Leg of Power Devices

In the previous section, we described a topology for the buck-boost converter which consists of two controllable power devices and two diodes. The converter has two separate modes of operation, one which results in the converter behaving as a buck converter and the other causing the converter to behave as a boost converter. The reader can modify the control file in the previous section to examine the behaviour of the converter in both modes. The converter presented in the previous section had one important characteristic—the two controllable power devices were distinct and separate devices. They will have to be purchased as two separate devices, and they will have their own separate gate drivers and other associated circuits such as protection circuits or snubbers. Moreover, the power flow in the converter was only possible in one direction—from the input to the output. In this section, we will describe how a power converter such as the one in the previous section can be realized in another manner which makes their packaging much easier and more efficient besides also allowing for flow of power in the reverse direction.

Let us begin with presenting a structure called the converter leg as shown in Fig. 3.7. Such a leg consists of potentially several controllable power devices connected in series. The left of Fig. 3.7 shows a two level structure while the right of Fig. 3.7 shows a three level structure. Intermediate connection points can be extracted as terminals besides the two extreme connection terminals. Each controllable device has a diode connected across it, but in such a way that the diode will conduct a current in a direction opposite to that of the controllable power device. For this reason, the diodes are called anti-parallel diodes. On the right of Fig. 3.7, another two diodes are usually used as shown by the box with dashed lines, and a later chapter will describe the operation of this topology in detail. Such structures are readily available from several manufacturers and they contain not only the controllable devices and the anti-parallel diodes, but also gate driver circuits, protection circuits and sometimes certain advanced functionalities [4, 5]. Therefore, when using multiple power devices, such modules are preferred over using distinct power devices, as it results in a more compact converter package.

Fig. 3.7 Converter legs—two level (left) and three level (right)

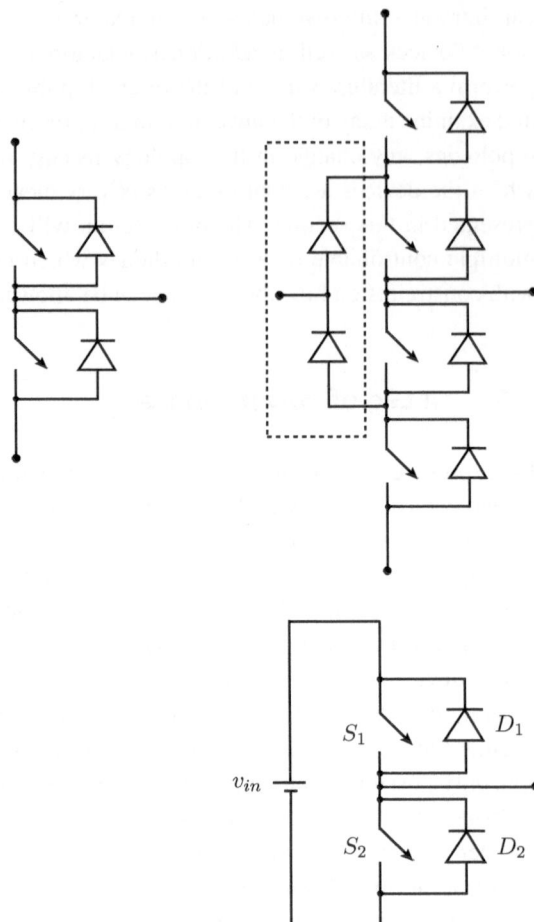

Fig. 3.8 Typical input voltage connection of the converter leg

Though in theory, we could use these modules in any way we see fit, with them being connected together as shown in Fig. 3.7, a few limitations and a few preferred approaches to their usage should now be discussed. In most applications that use such modules, the extreme terminals of the modules are typically used to form the dc bus. If we begin with the case of the buck converter, this would imply that the input dc voltage v_{in} will be connected across the extreme terminals and the mid-point of the leg extracted as an output terminal as shown in Fig. 3.8. Under such circumstances, there is one limitation to how the devices S_1 and S_2 that form the leg can be switched. Both devices cannot conduct at the same time, as in that case the dc bus will be short-circuited, and the large current that will flow can damage the devices or even the power supply itself. Though this might seem quite simple by ensuring that our controller does not result in trying to turn ON both controllable power devices S_1 and S_2, we also need to account for the fact that the devices need a finite amount of time to transition from cut-off to saturation and vice-versa.

Every manufacturer of controllable power devices provides detailed datasheets with the power devices which provides a number of important specifications of the device [4, 5]. A few are related to safety such as the maximum current that can flow through it, the maximum voltage that can be applied across it and a few others. Many others are related to the performance of the device under varying temperatures. A few of the specifications deal with the dynamic performance of the device. Within these specifications, one can find the minimum time needed for the device to turn ON and the minimum time needed for the device to turn OFF. When working with any power device, it is always advisable for a design engineer to be aware of all maximum ratings as well as some of the dynamic performance aspects of the device. When designing a prototype, one may need to use the temperature charts to determine the efficiency of the converter, design heat sinks and many more details. Without going into too much detail right in the beginning, let us begin our discussion with the turn-on time and the turn-off time of a power device.

Every controllable power device needs a finite amount of time to turn ON i.e transition from cut-off region to saturation region. This time interval can range from a few nanoseconds in very high speed silicon carbide MOSFETs to a few microseconds in high current rating IGBTs. This time interval is the time taken subsequent to a gate voltage being applied between either the gate-source terminals or the gate-emitter terminals, and the device being flooded with majority charge carriers and thereby going into saturation. A finite time is also needed for a device to turn OFF i.e transition from saturation to cut-off which again can range from a few nanoseconds to a few microseconds. This is the time interval needed after a negative gate-source or gate-emitter voltage is applied, which removes majority charge carriers from the device and re-establishes the barrier junctions. Therefore, when we turn OFF a device we must take into account that it will not stop conducting immediately, but only after a certain time interval, and until then the device continues to conduct.

In the simulations above, we have assumed ideal controllable devices. These devices can turn ON and turn OFF instantaneously. For most simulation studies, this assumption is quite acceptable unless one is simulating a very special case where one needs to examine how a converter behaves during the time period when a device is changing its conduction state. However, in a practical hardware implementation, one cannot make this assumption and it is necessary to introduce a blanking time. This blanking time is when we have turned OFF one controllable device and are waiting for the device to stop conducting, and until then we do not turn ON the other controllable device. This will introduce an interval during which no controllable device will conduct, as the blanking time is usually greater than the turn OFF time of a device, so as to ensure that it is completely turned OFF before the other device begins to turn ON.

Before the advent of microcontrollers and other digital controllers, such a blanking time was introduced using analog circuits. The reader can find several references to blanking time circuits in literature. In modern times, with the use of DSP based microcontrollers for implementing control loops and PWM, blanking time can be introduced using convenient settings in the PWM module of microcontrollers. Since this is a very hardware specific feature, this will not be covered in this book. However, for a reader who wishes to translate any of the switching strategies in this book to hardware prototypes, it is essential that he or she implement a blanking interval using the microcontroller being used for control and switching. In this book, since we are only dealing with simulations, ideal devices will continue to be used, and therefore, a blanking time need not be implemented. However, it is important to ensure that the two devices that form a leg should not conduct simultaneously at any point of time, and the gate signals provided to them must always be complimentary.

Before closing this discussion, let us address one issue which was mentioned before—the presence of the anti-parallel diodes. These diodes are connected across the controllable devices in a leg to ensure that current can still flow through the parallel combination in either direction. As an example, the controllable device can conduct current in only one direction (the downward direction). However, a current may be trying to flow in the upward direction. This usually happens when an inductor is connected to the mid-point terminal and the current flowing through an inductor is in such a direction that it tries to flow upwards through the device. If the anti-parallel diode was absent, the controllable device alone could not conduct a current in the opposite direction, and there would be no path for the inductor to flow. In designing any power electronic converter, it is important that inductor current should always have a path to flow, as breaking the current through an inductor abruptly can result in large induced emfs which can result in voltage spikes at different parts of the circuit. Therefore, these anti-parallel diodes allow the current through the parallel combination to flow in either direction, which in turn enables bidirectional power flow. This operation will become clear once we examine a few simulations.

In this section, we introduced the converter leg containing multiple power devices and their associated anti-parallel diodes. Though there are numerous benefits of such legs, the first and foremost advantage that will be evident to the reader is to make the converter significantly compact and efficient. This is due to the fact that now not only are the controllable power devices supplied by well-established companies after rigorous quality-assurance (QA) testing, but all the associated circuits such as gate drivers and protection circuits available in modules that contain one or more such legs, are also equally well tested. This results in a power converter will lesser unknown factors, and the general reliability of the system increases. In the next section, we will introduce through simulations, how the use of such a converter leg enables bidirectional flow of power in the converter.

3.4   Bidirectional Buck Converter

In the previous chapter, we had described the topology and operation of a buck converter along with a simulation. The topology in the previous chapter used a single discrete controllable power device and a single discrete diode, and will only allow flow of power from the input to the output. Therefore, if this buck converter were to be used for charging a 9 V battery from a 24 V dc source, it is not possible to reverse the application i.e charge a 24 V battery from a 9 V dc source. This is due to the fact that the converter is a buck converter and the output voltage will be lesser than the input voltage, and it is not possible for power to flow from the lower voltage to the higher voltage as the controllable power device will not allow current to flow in the reverse direction. In this section, we will examine how the buck converter of the previous chapter can be modified to a buck converter that allows bidirectional flow of power.

The modification to the buck converter so that it may allow bidirectional flow of power is quite simple once we precisely define what we expect from the converter. Let us suppose we have two levels of dc voltages—24 and 9 V, and we wish to have a power converter that enables us to supply power at 9 V dc if a 24 V dc power supply is available, as well as supply power at 24 V dc if a 9 V dc power supply were available. For this to happen, we need a buck converter between the 24 V dc and 9 V dc and a boost converter between the 9 V dc and the 24 V dc. These two circuits can be shown in Fig. 3.9, where the input voltage $v_{in} = 24$ V and the output voltage $v_o = 9$ V. The reader is encouraged to verify the buck and boost nature of the circuits from the past section which described the operation of the buck-boost converter with two discrete controllable power devices. Figure 3.10 shows the two operating states of the two converters. For both converters, there exists the third state where both the controllable power device and the diode are not conducting, and the inductor current is completely zero, but this has not been shown.

From Fig. 3.10, it is very clear that in order to achieve a transformation of voltage, we must combine the voltage of the power supply with the energy stored in the inductor. When the 24 V supply feeds 9 V, the 24 V dc provides energy to both the inductor as well as the 9V output, while when the 9 V supply feeds 24 V, the 9V dc and the energy in the inductor together feed the 24 V output. Moreover, in order to achieve bidirectional power flow, it is absolutely necessary that the current through the inductor be able to flow in both directions. We need to synthesize a single dc-dc converter that can achieve all the operating states as shown in Fig. 3.10. Figure 3.11 shows such a converter formed using a converter leg. The reader is encouraged to verify comparing with Fig. 3.10 that this converter in Fig. 3.11 will allow both buck and boost operations while also allowing flow of power in the reverse

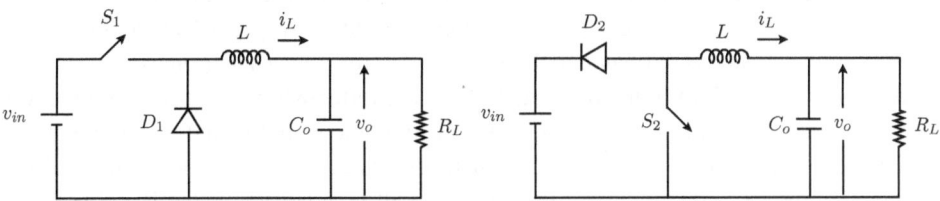

Fig. 3.9 Buck converter with forward power flow (left), boost converter with reverse power flow (right)

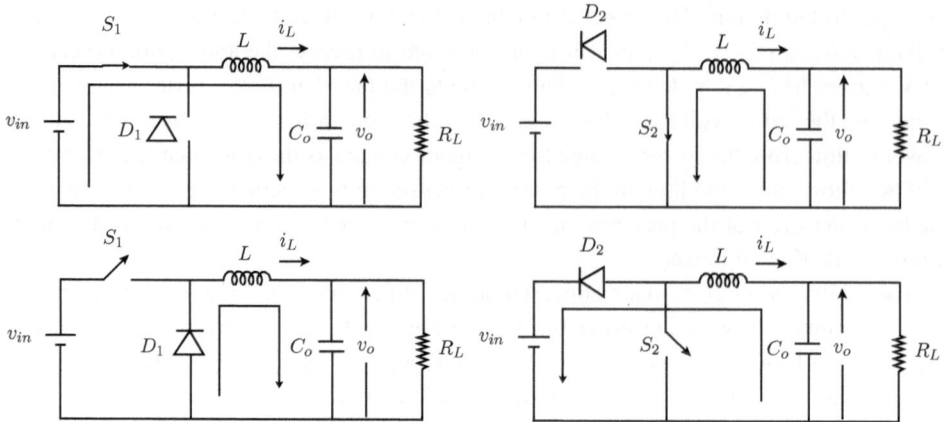

Fig. 3.10 Conduction states of the buck and boost converters when devices S_1 and S_2 are turned ON and OFF

Fig. 3.11 Bidirectional buck converter

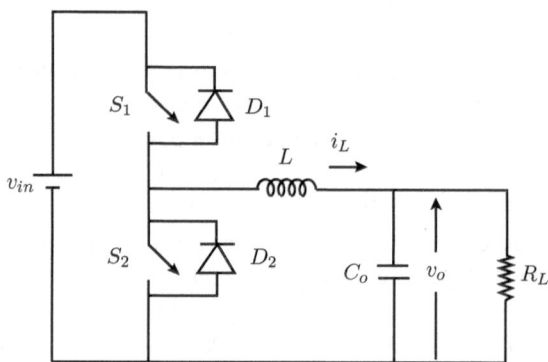

direction. For buck operation with forward power flow (v_{in} to v_o), the controllable power device S_1 and diode D_2 operate while the device S_2 is completely OFF. For operation with reverse power flow (v_o to v_{in}), the controllable power device S_2 and the diode D_1 operate while the device S_1 is completely OFF.

The simulation for the bidirectional buck converter can be found in the folder `bidirectional_buck_converter` within the folder `chapter3` at the link: https://github.com/opensourceelectrical/switching-strategies-for-power-electronics.

In order to verify with a simulation that the buck converter in Fig. 3.11 allows bidirectional flow of power, we will perform two independent simulations. In the first simulation, the 24 V dc will be a 24 V dc voltage source that feeds power to the 9 V dc which will be a capacitor. In the second simulation, the 24 V dc will be a capacitor that is fed power by the 9 V dc which is a 9 V dc voltage source. In the first simulation, the controllable power device S_2 is completely turned OFF and only device S_1 is given gating pulses. In the second

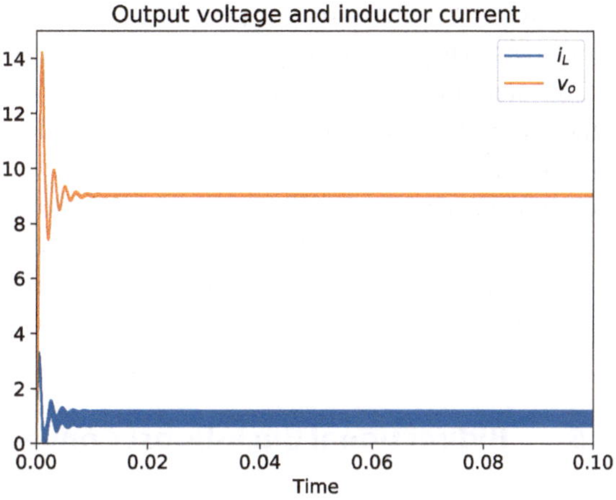

Fig. 3.12 Operation of buck converter with power flow in forward direction

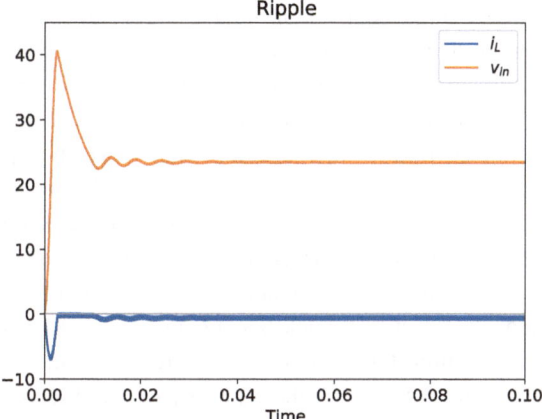

Fig. 3.13 Operation of buck converter with power flow in reverse direction

simulation, the controllable power device S_1 is completely turned OFF, and only device S_2 is given gating pulses. Figures 3.12 and 3.13 show the simulation results for the two cases. The reader is encouraged to plot all the results and zoom in to examine them in greater detail. However, only the most important results have been shown. In Fig. 3.12, it is clear that the inductor current in the direction shown in Fig. 3.11 is positive which implies that current is flowing from the 24 V supply to the output capacitor at 9 V. In Fig. 3.13, the inductor current is negative, which implies that an average current is flowing from the 9 V dc supply to the capacitor at 24 V. These results clearly show the bidirectional nature of the power converter. The reader can also examine the simulation of the bidirectional boost converter in folder `bidirectional_boost_converter` within the folder `chapter3` at the link: https://github.com/opensourceelectrical/switching-strategies-for-power-electronics.

In this section, we examined how a bidirectional buck converter can be synthesized using a converter leg with two controllable power devices and their anti-parallel diodes. With this converter, we could have two power supplies at the two ends of the converter, and allow them to exchange power with each other. Such a converter could prove useful if we wish to charge and discharge a battery from a source that could at times supply power and at other times could benefit from receiving power. Without going into significant details, one can think of the battery of an electric vehicle, which would normally supply power to a motor so that the car can move as the motor rotates, but can also receive energy when the car brakes in which case the mechanical energy of the motor is converted back into electrical energy in a process called regenerative braking. In the next section, we will examine how a bidirectional buck-boost converter can be synthesized using two converter legs.

3.5   Bidirectional Buck-Boost Converter

In the previous section, we had examined the operation of a bidirectional buck converter where a single controllable device and a single freewheel diode were replaced by a converter leg consisting of two controllable power devices with their associated anti-parallel diodes. At first glance, it might appear that not much has been gained, as instead of one controllable power device and a diode, we now use two controllable power devices and two diodes. However, besides the significant advantage that power can now flow in both directions, the controllable devices and diodes can be inserted into the circuit as a single module commercially available from manufacturers of components, rather than have a separate controllable device and diode as discrete components. This greatly decreases the size of the converter and makes it much more convenient to package. In this section, we will examine how a bidirectional buck-boost converter [6–8] can be realized using two converter legs in an operation strategy not very different from the previous sections.

Figure 3.14 shows the topology of a bidirectional buck-boost converter [6–8]. As can be seen, both the input v_{in} and the output v_o are interfaced through converter legs. The inductor is connected between the mid-points of the two converter legs, and exchanges energy between the input and the output. For a forward buck operation, where the output voltage v_o is lesser than the input v_{in} and power flows from v_{in} to v_o, the controllable device S_1 operates and diodes D_2 and D_3 support conduction. When S_1 conducts, the energy is stored in the inductor and also fed to the output through the diode D_3, while when S_1 stops conducting, the energy in the inductor freewheels through the diodes D_2 and D_3 while supplying the output capacitor. For a reverse buck operation, where the output v_o is lesser than v_{in}, but power flows from the output v_o to the input v_{in}, the controllable device S_3 continuously conducts while device S_2 is switched intermittently. When S_2 and S_3 conduct, energy is stored in the inductor, and when S_2 is turned OFF, the energy in the inductor freewheels through the diode D_1 and device S_3. For a forward boost operation, with output v_o being greater than input v_{in} and power flowing from v_{in} to v_o, the controllable device S_1 operates continuously, while S_4 is switched intermittently. When S_1 and S_4 are both conducting, energy is stored in the inductor, and when S_4 stops conducting, the energy in

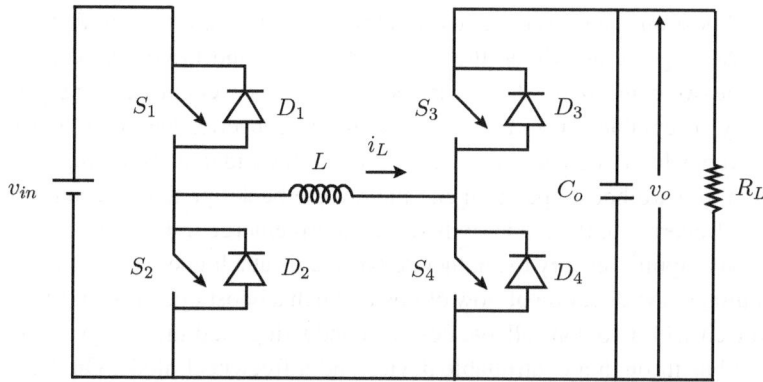

Fig. 3.14 Bidirectional buck-boost converter

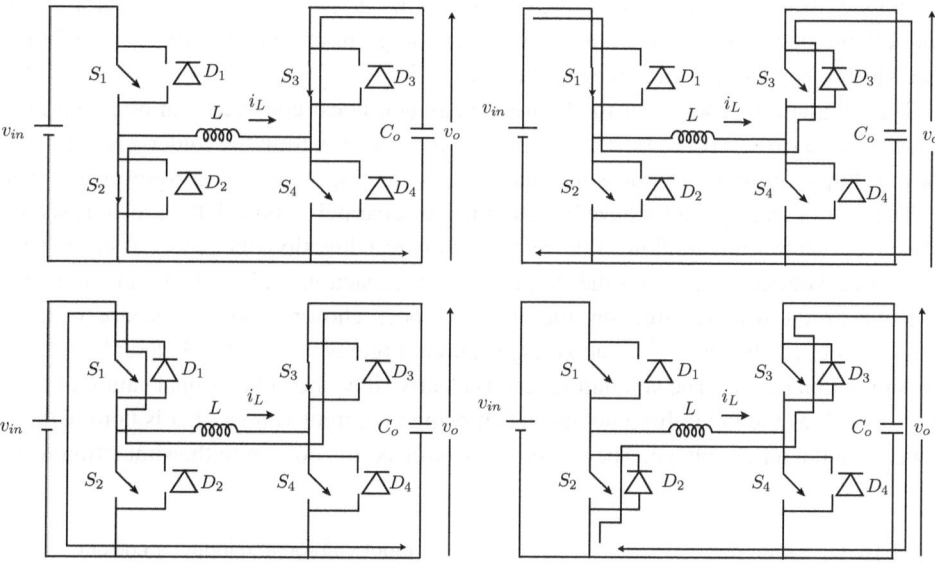

Fig. 3.15 Bidirectional buck-boost converter operation - reverse buck (left) and forward buck (right)

the inductor freewheels through the diode D_3 and S_1. For a reverse boost operation, with output v_o being greater than input v_{in} and power flowing from v_o to v_{in}, the controllable device S_3 operates while diodes D_1 and D_4 support conduction. When S_3 conducts, the energy is stored in the inductor while the input receives power through the diode D_1, and when S_3 stops conducting, the energy in the inductor freewheels through diodes D_1 and D_4 and continues to supply energy to the input.

Figure 3.15 shows reverse buck and forward buck operation of the converter. The reader is encouraged to draw the operating states of the remaining two modes as described above—

the forward boost and the reverse boost modes. It is important to note that since we are using converter legs across which the input voltage v_{in} and the output capacitor C_o are connected, the two controllable devices in a single leg are never conducting simultaneously. It is perfectly acceptable for an upper device in one converter leg and a lower device in another converter leg to conduct, as in such a mode the inductor is storing energy, which occurs in the forward boost operation and the reverse buck operation. Upon inspecting all four modes of operation, the reader will see a pattern emerge in how the devices conduct. When a dc bus supplies energy to another dc bus that is at a higher voltage, this results in a boost operation in the direction of flow of power. Such a boost operation needs the inductor to be energized with two controllable devices conducting, and the energy to be fed to the receiving dc bus through a controllable device and a freewheel diode. On the other hand, when a dc bus supplies energy to another dc bus that is at a lower voltage, this results in a buck operation in the direction of flow of power. Such a buck operation needs energy to be supplied from the sending dc bus to the receiving dc bus through the inductor using a controllable power device and a diode, and the energy in the inductor to freewheel to the receiving dc bus through two freewheel diodes.

The simulation package for the bidirectional buck-boost converter can be found in the folder `bidirectional_buck_boost_converter` within the folder `chapter3` at the link: https://github.com/opensourceelectrical/switching-strategies-for-power-electronics.

The four operating modes have been simulated separately. As with the previous section, when we wish to simulate flow of power in the forward direction, the input has been chosen to be a dc voltage source and the output to be a capacitor, while when simulating flow of power in the reverse direction, the input has been chosen to be a capacitor while the output has been chosen to be a dc voltage source. Figures 3.16, 3.17, 3.18 and 3.19 show the simulation results. The simulation results clearly show the bidirectional nature of power flow as well as the flexibility one has in achieving an output voltage that is both higher or lower than the input voltage. The reader is encouraged to zoom in to the simulation results

Fig. 3.16 Operation of buck-boost converter with power flow in reverse buck mode

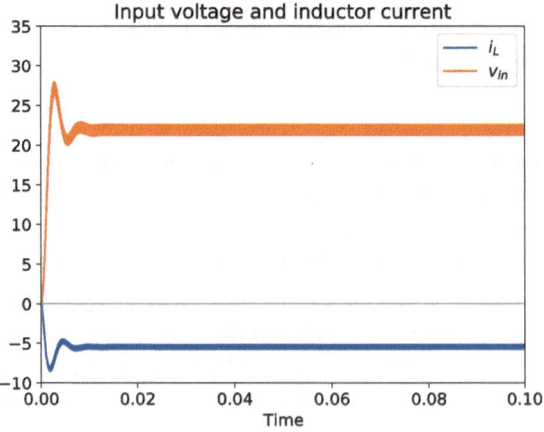

Fig. 3.17 Operation of buck-boost converter with power flow in reverse boost mode

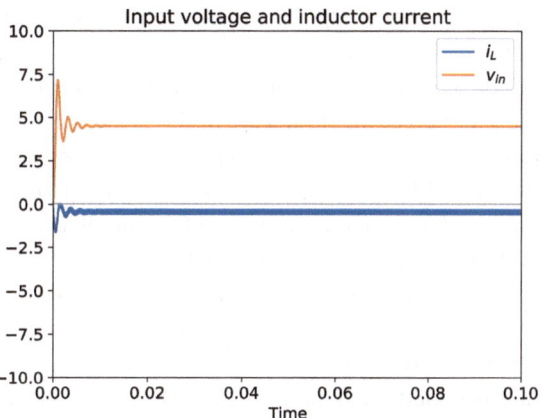

Fig. 3.18 Operation of buck-boost converter with power flow in forward boost mode

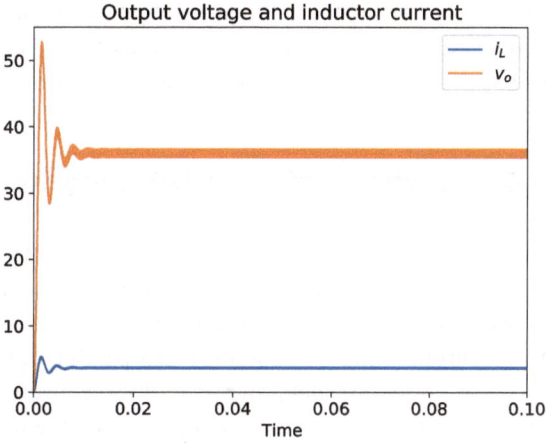

Fig. 3.19 Operation of buck-boost converter with power flow in forward buck mode

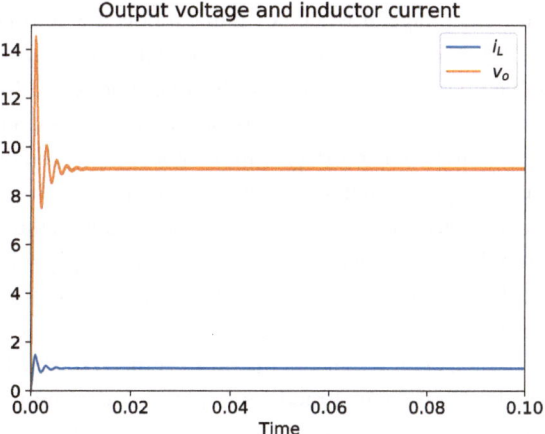

and compare the ripple in the inductor voltage with the ripple in either the output capacitor or the input capacitor depending on the nature of power flow. The reader is also encouraged to note that in the control code, when a controllable device in any given leg is turned ON, the other controllable device in the same leg has been explicitly turned OFF. When one translates control code into hardware, one needs to ensure that sufficient blanking time is introduced between the switching of controllable devices in a leg in addition to ensuring that the devices are never turned ON simultaneously.

In this section, we have simulated a bidirectional buck-boost converter. This converter is an extension of the bidirectional buck converter presented in the previous section, and offers greater flexibility over interfacing a particular input voltage to an appliance at another voltage. This flexibility is very important, as quite often power supplies are not standardized internationally, and one can find different voltage levels as one travels. Therefore, it is quite important that our power supplies work for a range of voltages and not just a single voltage. In the next section, we will present a quick summary of the dc-dc converters described and simulated, so as to describe the importance of taking into consideration the basic principles of physics while designing power electronic converters.

3.6   Summary of dc-dc Switching Strategies

In the previous chapter, we had described the operation of a conventional buck converter, while in this chapter, we had described the operation of modified buck and buck-boost converters synthesized using discrete devices as well as with converter legs consisting of two controllable devices and their anti-parallel diodes. Before we conclude our discussion on dc-dc converters and progress to dc-ac converters in the next section and also the later chapters, let us draw a few summaries from the topologies that we had examined. Such a summary will also highlight some of the challenges and also the flexibilities that one can take advantage of when designing a power electronic converter.

When examining the modified buck-boost converter comprised of two discrete controllable power devices and two discrete diodes, we found two distinct modes of operation of the converter. In the buck mode, device S_2 was completely turned OFF while device S_1 was switched using PWM. In the boost mode, device S_1 was completely turned ON, while device S_2 was switched using PWM. In the boost mode, it was essential for both controllable devices to conduct simultaneously, due to the fact that in boost mode energy was being stored in the inductor from the input dc supply with both controllable devices conducting. In this topology, there was no forbidden mode of operation, and it would not be necessary to check for a particular state of conduction of the devices, and prevent it from occurring.

When using a converter leg for synthesizing the buck or buck-boost converters, we examined several distinct modes of operation depending on what we wished to achieve—either buck or boost operation, and forward or reverse power flow. In the topologies we examined, either a dc voltage source or a capacitor (electrolytic) that forms a dc bus was connected across the converter leg. In such a case, a forbidden mode of operation was for both controllable devices to conduct simultaneously. Towards this end, in every PWM code, when a controllable device in a leg was turned ON, the other controllable device in the same leg was turned OFF. Ensuring that the gate pulses for the controllable devices are complimentary is sufficient in a simulation. However, when implementing this PWM strategy in hardware, it is also essential to implement a blanking time, when no gate pulses are provided to the controllable devices in a leg to ensure that the device that is turning OFF completely stops conducting before the other controllable device is turned ON.

While examining different modes of operation, we had examined the operation of the converter only in open loop without any voltage or current regulation strategy. In any practical converter, a closed loop strategy is essential to ensure that the output voltage remains close to the desired value. Furthermore, there might be other control objectives such as ensuring a certain amount of current flowing between the input and output, or limiting this current to a certain maximum safe value. Control varies vastly with the application, and while one application might require the output voltage to be regulated very close to a reference, another application might need the current to be regulated to a certain maximum while the output voltage need only be reasonably close to the reference. As an example, for a dc-dc converter used as a power supply, the output voltage is of utmost importance as it will supply a consumer appliance, while for a dc-dc converter used in an electric vehicle, the current might need to be regulated to ensure that the desired motor torque is produced.

In our description of the operation of dc-dc converters, particularly with the case of bidirectional buck and bidirectional buck-boost converters, we had divided their operation into distinct modes. Such a division made understanding their operation much simpler, and also the resultant simulations needed simpler PWM generation control files. However, for the sake of completing our discussion on these converters, but still not dealing with the complexities of closed-loop control, let us consider the case where the operating mode of a converter will transition from one mode to another. As an example, let us consider the case where the operation mode of the bidirectional buck-boost converter changes from forward buck mode to reverse buck mode. Such a transition might occur when it becomes necessary for the output to supply energy back to the input, which might for example occur during regenerative braking of an electric motor, though the output voltage and the input voltage may not change significantly during the transition.

In the forward buck mode, device S_1 operates along with diode D_3 while diodes D_2 and D_3 conduct during freewheeling of the inductor energy. In the reverse buck mode, devices S_2 and S_3 operate to store energy in the inductor from the output, while the inductor energy and output feed the input through S_3 and diode D_1. When transitioning from one mode of operation to another, it is usually necessary to ensure through control loops that the transition is smooth and without spikes in either the current or any of the voltages. However, if we ignore control strategies, and merely change our mode of operation abruptly, let us examine through analysis as well as simulation what the results would be. Figure 3.20 shows the reverse buck mode of operation under two conditions—the first (left) when the current still

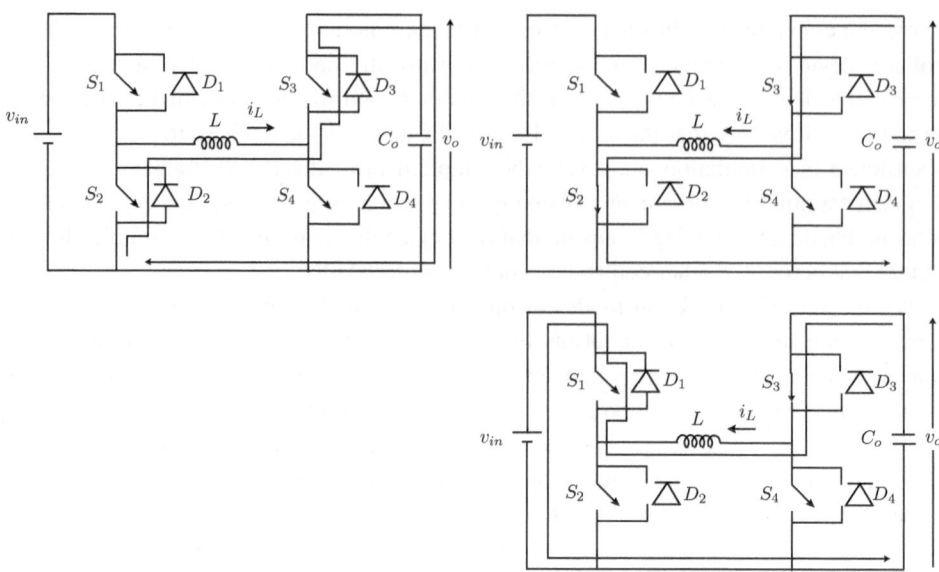

Fig. 3.20 Bidirectional buck-boost—transition from forward buck (left) to reverse buck (right)

flows from the input to the output and the second (right) when the current has reversed and flows from the output to the input. Though we have changed the operation of the controllable power devices, the current through the inductor will not reverse abruptly and this current will gradually reduce to zero and then increase but in the negative sense.

Though S_2 and S_3 will be given gate pulses with the intention of turning them ON, due to the direction of the inductor current, they cannot conduct. The only path that the current can flow in is through the diodes D_2 and D_3 which remains one of the states of the forward buck mode of operation. This is an example of how even though we wish to turn ON controllable devices so as to energize the inductor, due to the existing direction of inductor current, the devices cannot conduct and the anti-parallel diodes across the devices end up conducting instead. Similarly, diode D_1 also will not be able to conduct a current in the reverse direction, and therefore, turning S_2 OFF will have no effect as the current can only flow through D_2 and D_3. This state will continue until the inductor current becomes zero and diodes D_2 and D_3 will stop conducting. Only after the current becomes zero, that turning ON S_2 and S_3 will have any effect as now the inductor current will reverse and the energy will be stored in the inductor. When S_2 is turned OFF, the output and the energy stored in the inductor will flow through controllable device S_3 and diode D_3. The converter will then completely transition to reverse buck mode of operation.

It is immediately evident that by abruptly transitioning from a forward buck mode to a reverse buck mode of operation, we have no control over the converter for a certain period of time while the energy in the inductor freewheels and dissipates. How fast the

Fig. 3.21 Buck-boost
converter inductor current
during transition from forward
buck mode to reverse buck
mode

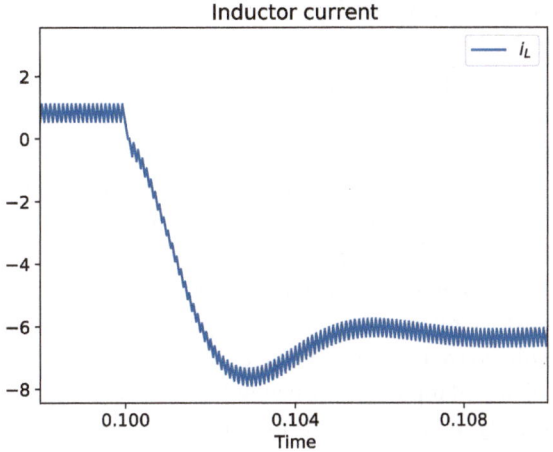

inductor current falls to zero is decided by the load connected across the output capacitor. Normally one would not like the operating state of a converter to be decided by external factors such as the load connected, as this can always change instantaneously. Therefore, this is a limitation of the converter as it offers no alternative path by which the current through the inductor can be forced to decrease to zero at a faster rate. The simulation of this case can be found in the folder `operating_mode_transition` in folder `bidirectional_buck_boost_converter` within the folder `chapter3` at the link: https://github.com/opensourceelectrical/switching-strategies-for-power-electronics.

Figure 3.21 shows the simulation result in which the inductor current falls to zero and then reverses when the operating mode changes from forward buck to reverse buck. In the simulation, this has been achieved by changing the switching states at the time instant of 0.1 s after the start of the simulation.

The reader is encouraged to try out the transition between other modes of operation as well as try and simulate them. In this section, we summarized the operation of the various dc-dc converters that have been examined so far. It is important when synthesizing a power electronic converter, that the design engineer marks out the preferred, the allowed and the forbidden states of conduction. At this point, the reader might not be able to distinguish between the preferred and the allowed conduction states, as the converters examined so far have distinct modes of operation with little or no degree of freedom. In the later chapters, this distinction will become clear, as the reader will immediately observe that for a particular mode of operation, there are many possibilities in terms of switching controllable devices. However, it is essential that the switching strategy should ensure that forbidden states do not occur. Until this section, all our examples are related to purely dc systems. However, it is possible to switch the controllable devices in a leg in any manner, and therefore, these converter legs can also be used in ac systems as will be shown in the next section and also the later chapters.

3.7   Half Bridge dc-ac Converter

In the previous sections, we introduced how a converter leg can be used to synthesize a number of converters—buck, boost and buck-boost, while also enabling bidirectional flow of power between the input and the output. However, these converters interfaced one dc voltage at a particular level to another dc voltage at another level. Though this application of power converters is vast, particularly for consumer electronics such as mobiles, laptops and many other household appliances, there is another very important application of converters—in interfacing a dc voltage and an ac voltage. In this section, we will get started with a very basic application, where an appliance can be supplied by an ac current when the input voltage source is a dc voltage. This converter, however, only opens the door to all the other advanced topologies which will follow in the later chapters. The objective of this section is merely to highlight how useful this converter leg can be in power electronics, and how, for power electronics, this converter leg is a building block to synthesizing complex converter topologies.

Before we describe the topology of such a dc-ac converter, let us look very briefly into an application of a dc-ac inverter, though not necessarily the converter topology that we are going to describe in this section. In many cases, the voltage available from energy sources is an unregulated dc voltage. For example, a solar photovoltaic (PV) panel produces a dc voltage whose value changes with the incident solar radiation. A 24 V dc battery such as the ones used in the simulations of the previous sections will also produce a variable dc voltage whose value will change as the state of charge of the battery changes i.e the battery charges or discharges. Such variable dc voltage sources will need to supply power to ac systems in certain applications such as grid integration of solar PV or electric vehicles that use ac motors such as induction motors or synchronous motors. This can be made possible by the use of dc-ac converters which are also commercially called inverters. Most of the remainder of this book will examine many different topologies of dc-ac converters and the switching strategies that can be implemented with them.

In the previous simulations where a converter transformed a dc voltage from one level to another, it was sufficient to produce a switched voltage which could be filtered to produce an output with a tolerable amount of ripple. In the case of bidirectional converters, the current could flow in both directions, resulting in an average current that had a positive or a negative value. However, in all the simulations, the required output voltage was always a dc voltage and as a result the currents in different parts of the circuit had an average dc value. The ripples that were observed in either the output voltage or the inductor currents, were merely due to the switching of the controllable power devices and were superimposed on the average dc values of the output voltage and the inductor current. When interfacing a dc voltage to an ac system, we now have to produce an ac voltage from the dc input so as to either supply this ac voltage directly to an appliance (for example an electric vehicle motor) or to generate an ac current (to feed power into the grid in the case of solar PV).

In the previous chapter, we had very briefly examined how an ac voltage can be synthesized from a dc voltage while describing the process of Pulse Width Modulation (PWM). In fact, we had used this application to describe the benefits of using PWM versus producing a rectangular waveform that contained a fundamental ac waveform. The PWM is achieved by comparing a modulation signal with a carrier waveform. The modulation signal is a sine waveform of the same frequency as the ac waveform that is desired, while the carrier waveform is a high-frequency triangular waveform. This triangular waveform produces a switching pulse which is symmetrical over a switching cycle and differs from the sawtooth waveform used in the previous simulations of dc-dc converters. The reader is encouraged to review the sections "Introducing modulation in power electronics" and "Pulse Width Modulation (PWM)" in the previous chapter.

Figure 3.22 shows the topology of a very simple dc-ac converter called a half-bridge converter. The converter leg has the same structure as used before—two controllable power devices with associated anti-parallel diodes. The input dc voltage is connected across the extreme terminals of the converter leg and the mid-point of the converter leg is connected to the ac system. The ac system in our case is merely a inductor-resistor combination that could represent any random load. One could theoretically also connect a single-phase ac grid, however, including an independent ac grid will need additional control loops which is not the focus of this book. In order to complete the ac circuit, the neutral of the ac system is connected to the mid-point of a dc bus comprised of two electrolytic capacitors which are also connected across the dc input voltage. It is important to note that unlike the dc-dc converters in the previous simulations where the output and input shared a common ground, in this case of the dc-ac converter, the neutral of the ac system cannot be directly connected to the ground of the input dc voltage. To allow the circuit to complete, this neutral path needs to be separately formed through the mid-point of the dc bus capacitors [9]. In the next chapter, we will begin to look into full-bridge converters where the output terminals

Fig. 3.22 Half bridge dc-ac converter

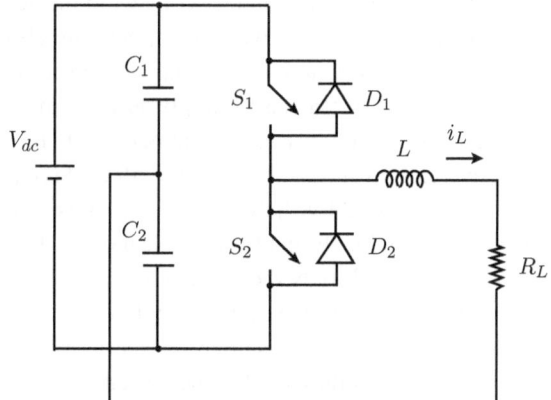

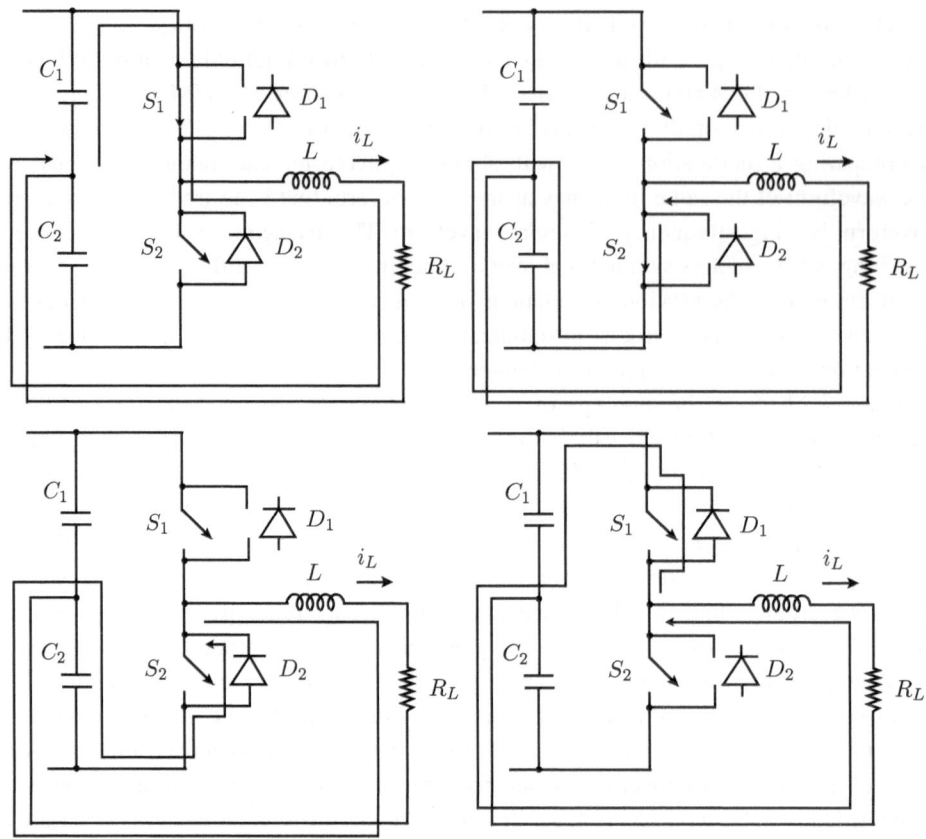

Fig. 3.23 Half bridge dc-ac converter operation—positive mode (left) and negative mode (right)

of the converter legs can directly connect to the ac system. In contrast, Fig. 3.22 is called a half-bridge as it does not have two output terminals for a complete interface to the ac system.

To explain the operation of this half-bridge converter, one can use several approaches [8]. The most direct approach is to simply describe the different operating states of the converter. Since the converter feeds an ac system, it would be simplest to divide the operating modes of the converter into one for the positive half cycle (let us call it positive mode) of the current i_L and one for the negative half cycle of i_L (let us call it negative mode). These two modes are shown in Fig. 3.23. During the positive mode, when the current i_L is flowing in the direction shown in Fig. 3.23, only the devices S_1 and D_2 can conduct while the device S_2 and diode D_1 cannot conduct as the current would be in the reverse direction. When the device S_1 is conducting, the dc voltage source $\frac{V_{dc}}{2}$ acts as a driving force and the current i_L will increase. When device S_1 is turned OFF, the current i_L will freewheel through the diode D_2 and the current faces a negative dc voltage source $-\frac{V_{dc}}{2}$ which acts as a retarding force and causes the current to decrease. During the negative mode, when S_2 is conducting, the negative dc

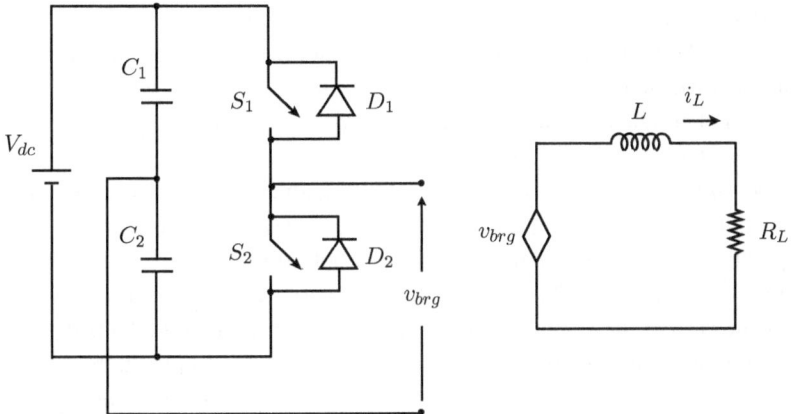

Fig. 3.24 Half bridge dc-ac converter equivalent circuit

voltage source $-\frac{V_{dc}}{2}$ acts as a driving force and causes the current to increase but in the negative sense. When S_2 is turned OFF, the current freewheels through diode D_1 and faces an opposing positive dc voltage source $\frac{V_{dc}}{2}$ which acts as a retarding force and causes the current to decrease in magnitude.

From the above discussion, we can simplify the half-bridge circuit to a simpler equivalent circuit as shown in Fig. 3.24. The dc voltage source, the dc capacitors and the converter leg have been represented as a controllable voltage source V_{brg}. This is possible if we consider the mid-point of the converter leg and the mid-point of the dc bus capacitors (also the neutral) to be the output terminals of the converter. When device S_1 or diode D_1 conducts, the controllable voltage source will produce a voltage $\frac{V_{dc}}{2}$, while when device S_2 or diode D_2 conducts, the controllable voltage source will produce a voltage $-\frac{V_{dc}}{2}$. The reader is encouraged to verify this with respect to Fig. 3.23. For such an equivalent circuit, the equation for the current is:

$$k\frac{V_{dc}}{2} - L\frac{di_L}{dt} - R_L i_L = 0 \qquad (3.10)$$

The reader can refer to the section Pulse Width Modulation in the previous chapter to review the sine-triangle approach to PWM for synthesizing an ac voltage waveform from a dc voltage source. In the previous chapter, we had merely used a switch controlled by PWM to synthesize a sine waveform and using DFT had determined that the switched waveform contained a frequency component equal to the frequency of the modulation signal and high frequency components that were clustered around multiples of the switching frequency. We can use this sine-triangle approach for the half-bridge dc-ac converter, such that for a modulation signal $m(t)$ and a carrier signal $c(t)$:

When $m(t) > c(t)$, S_1 is ON, S_2 is OFF and $k = +1$

When $m(t) < c(t)$, S_1 is OFF, S_2 is ON and $k = -1$

The reader must note that when one controllable device in the converter leg is turned ON, the other is turned OFF to ensure that the dc voltage source is not short-circuited. Moreover, during the positive mode, when S_2 is turned ON for the periods when $m(t) < c(t)$, the device S_2 cannot conduct because the inductor current is positive, and therefore, diode D_2 will conduct instead. Conversely, during the negative mode, when S_1 is turned ON for the periods when $m(t) > c(t)$, the device S_1 cannot conduct because the inductor current is negative, and therefore, diode D_1 will conduct instead. Therefore, even in a dc-ac converter, the anti-parallel diodes enable current to flow in both directions in a controllable device-diode parallel combination. This one constraint is very important in any power electronic converter - the current through an inductor must always be able to find a path to flow.

The simulation of the half-bridge dc-ac converter can be found in the folder `half_bridge_dc_ac_converter` within the folder `chapter3` at the link: https://github.com/opensourceelectrical/switching-strategies-for-power-electronics.

Figure 3.25 shows the waveform of the current for a narrow time window that covers a region in the positive half-cycle and the negative half-cycle. Figure 3.26 shows the currents i_{c1} and i_{c2} through the dc bus capacitors from which it is quite clear that the entire output current is flowing into the mid-point of the dc bus. The reader is encouraged to verify by zooming into the plots of how the current will rise and fall according to the description above when the devices are turned ON and OFF. Using sine-triangle PWM, we are able to produce an ac current having a sine waveform with a ripple using a dc voltage source. The reader is encouraged to change the phase angle of the modulation signal and examine the resultant current waveform. The reader should also note that the waveform of the current will not have the same phase angle as that of the modulation signal. This is due to the fact that the equivalent converter voltage feeds a resistor-inductor ac circuit.

The reader should note that while the half-bridge dc-ac converter produces an ac current using a dc voltage source and a single converter leg, due to the neutral of the ac circuit being

Fig. 3.25 Half bridge converter output current

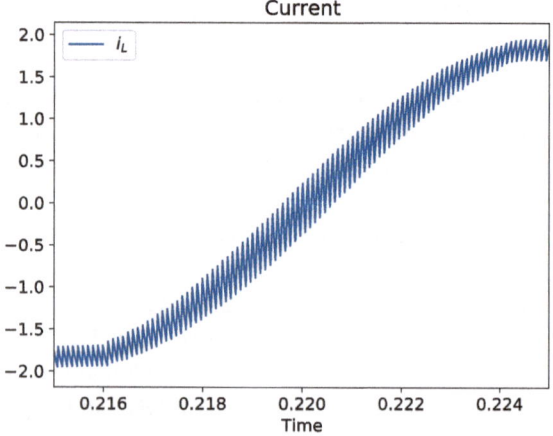

Fig. 3.26 Half bridge converter dc bus capacitor currents

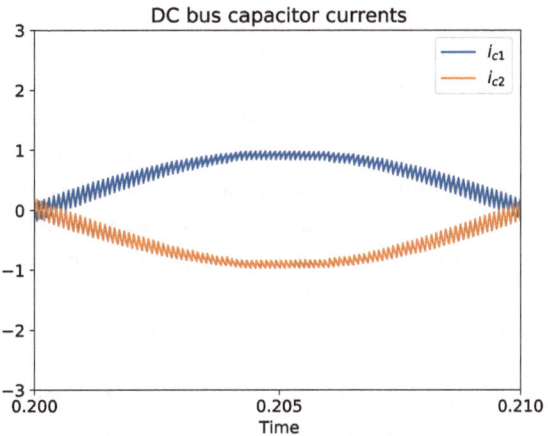

connected to the mid-point of the dc bus capacitors, the entire ac current will flow through the dc bus capacitors. Therefore, if the ac current was significantly large, the ripple in the voltage of the dc bus capacitors will also be large and if allowed to exceed a certain limit, can result in distortions in the ac current. This is due to the fact that the converter equivalent circuit assumes an output of either $+\frac{V_{dc}}{2}$ or $-\frac{V_{dc}}{2}$, and this assumption will no longer hold if the dc bus capacitor voltages were to have significant ripple. In the later chapters, many other topologies will be presented for dc-ac converters.

In this section, we examined how using a single converter leg, it is possible to realize a rudimentary dc-ac converter. The purpose of this section is to show how it is possible to synthesize power converters as long as one accounts for all modes of operation that would be needed for an application. The reader is encouraged to use these basic tools of synthesis to think of other potential ways in which a dc-ac converter can be synthesized using just the minimum number of controllable devices and diodes. There are a number of such circuits, though with limited application, but can still be very interesting to someone beginning to learn power electronics. Such an approach enables a learner to learn by doing rather than simply be presented circuits with a description of their operation.

3.8   Conclusions

In this chapter, we examined a number of power converters that used two controllable devices. We began the chapter by presenting a modified buck-boost converter which as the reader can observe is merely an extension of the buck converter described and simulated in the previous chapter. Interestingly, the converter can be operated separately as a buck converter or as a boost converter. In the buck mode of operation, only one controllable device intermittently conducts along with one diode. The other controllable device cannot conduct during buck mode of operation. During boost mode of operation, one controllable device conducts continuously, while the controllable device which did not conduct during buck mode now conducts intermittently along with another diode. Moreover, the converter needs two diodes to operate. Though one might be quick to jump to the conclusion that every controllable device must be accompanied by a diode, that is not a correct conclusion. A diode is connected to allow an inductor current to continue to flow when no controllable device is conducting, and also to prevent current flowing in a particular path.

The chapter then introduced the concept of a converter leg. For most of us power electronics engineers, the converter leg is a basic building block similar to the how elements are to a chemist for synthesizing compounds. Most power electronics texts will introduce power converters that have these converter legs without introducing their advantages and benefits. Converter legs consisting of controllable devices and diodes connected in various combinations are available commercially from many manufacturers as compact modules. These modules have multiple devices readily connected in several popular topologies making it easy to build converters with them as many connections are already formed internally. Moreover, these modules have associated circuity such as gate drivers and protection circuits such as overcurrent and overvoltage protection which make them very convenient to use and reduces the size and complexity of power converters formed using them.

The chapter describes how a buck and buck-boost converter can be designed using these modules. In addition to the regular buck and buck-boost operating modes, the use of these converter legs enables power flow in the reverse direction. This has been demonstrated in detail using circuit diagrams and also through simulation studies. The flexibility that is achieved through enabling bidirectional flow of power makes them extremely useful in many applications such as battery charging or in electric vehicles where a bidirectional flow of power is imperative. The use of converter legs does however introduce a limitation - all the controllable devices in a single leg cannot be conducting simultaneously, as this will result in a short-circuit of the dc voltage source connected across the converter leg. This limitation can be dealt with quite conveniently in simulations by ensuring that when one device in a converter leg conducts, the other is explicitly turned OFF. However, if the readers were to build a hardware prototype using a converter leg, they would need to additionally insert a blanking time which implies that gate signals will not be provided to either of the controllable devices for a time interval called the blanking time interval. This ensures that one controllable device will completely turn OFF before the other begins to conduct.

The chapter concludes with a simulation of a half-bridge dc-ac converter consisting of single converter leg. Such a converter is not widely used due to its limited current sourcing capability, but is a topology that demonstrates the usefulness of the converter leg. Using circuit diagrams and simulation results, the operation of the converter is described in detail. The simulation result shows how an ac current is produced by the converter. The reader is encouraged to examine other waveforms such as the gate signals provided to the controllable devices, as well as the voltage across the dc bus capacitors, in order to fully understand the working of the converter. In the next chapter, the reader will become familiar with the full-bridge topology which merely consists of two converter legs instead of one.

References

1. J. Chen, D. Maksimovic, R. Erickson, Buck-boost PWM converters having two independently controlled switches, in *IEEE 32nd Annual Power Electronics Specialists Conference (IEEE Cat. No. 01CH37230)*, vol. 2. (IEEE, 2001), pp. 736–741
2. N. Mohan, T.M. Undeland, W.P. Robbins, *Power Electronics: Converters, Applications, and Design*, 3rd edn. (Wiley, New York, 2002)
3. R.W. Erickson, D. Maksimovic, *Fundamentals of Power Electronics* (Springer Science & Business Media, 2007)
4. *5SNG 0600R120500 LoPak IGBT module* (Hitachi Energy, 2022)
5. *5SNG 0250P330305 HiPak IGBT Module* (ABB, 2019)
6. R. Sosnowski, M. Chojowski, M. Baszyński, An analysis of a transformerless dual active half-bridge converter. Power Electron. Drives **7**(1), 146–158 (2022). [Online]. Available: https://doi.org/10.2478/pead-2022-0011
7. *4-Switch Buck-Boost Bi-directional DC-DC Converter Reference Design* (Texas Instruments, 2021)
8. N. Su, D. Xu, M. Chen, J. Tao, Study of bi-directional buck-boost converter with different control methods, in *IEEE Vehicle Power and Propulsion Conference* (IEEE, 2008), pp. 1–5
9. M.H. Rashid, *Power Electronics: Circuits, Devices, and Applications* (Pearson Education India, 2009)

Full-Bridge Converter

4

4.1 Introduction

In the previous chapter, we examined power converters with two controllable power devices, and thereby introduced the concept of coordinated switching of controllable devices. Subsequently, we introduced the single converter leg comprising of two controllable devices and their associated anti-parallel diodes, and examined how such a converter leg can be used a fundamental building block for other power converters. We used simple permutations and combinations to examine the different switching combinations possible, and also were able to conclude that it would not be permitted to turn ON both controllable devices in the leg to avoid short-circuiting the dc bus. Using the examples of buck, boost, buck-boost and the half-bridge converter, we examined different conduction modes of the converters and devised PWM strategies in all cases.

In this chapter, we will extend our discussion in the previous chapter by presenting the full-bridge converter which is merely a combination of two converter legs. Similar to the single converter leg, the full-bridge converter is also extremely popular and is commercially available as a module from many manufacturers. One can find numerous applications of the full-bridge converter in both dc-dc and dc-ac applications, and therefore, this chapter will describe through analysis and simulations several applications. One can find this converter dealt with in many textbooks and research articles, and therefore, it is necessary to describe, what might be considered to be novel in this chapter with respect to other literature. Similar to the previous chapters, this chapter will not directly jump into switching strategies for the full-bridge converter, and subsequently examine practical applications. This chapter will instead use another approach to make the matter easier to comprehend for a newcomer to power electronics, while also providing a few tools and techniques to both understand and devise PWM strategies.

The chapter will begin with examining the different possible conduction modes of the full-bridge converter, as this should be the first step towards getting acquainted with any

S. V. Iyer and M. N. Aalam, *Switching Strategies for Power Electronic Converters*,
Synthesis Lectures on Power Electronics, https://doi.org/10.1007/978-3-031-41405-3_4

new converter topology. Due to the presence of two converter legs, which in turn implies four controllable devices and their associated anti-parallel diodes, the number of conduction modes will be greater than the single converter leg presented in the last chapter. A few of the conduction modes result in similar conditions as compared to the half-bridge converter, and it is possible to devise a PWM strategy to use only these conduction modes. Before directly examining PWM strategies to utilize the other conduction modes, the chapter will introduce the concept of how the converter voltage can be expressed as a vector that can attain different values. With such a vector representation of the converter voltage, it becomes quite simple to think of a switched voltage pattern with the different possible voltage levels and the equivalent vector transition to achieve this switched voltage pattern.

The chapter will then examine a few popular PWM strategies and the switched voltage pattern that results, and examine how these can be achieved using vector transitions. Furthermore, the chapter will also examine how for any given desired switched voltage pattern, it is possible to synthesize carrier waveforms and algorithms to compare them with duty ratios or modulation signals. The algorithms for generating carrier waveforms and the comparison logic will be developed in an iterative manner such that the reader will be able to develop a technique for adapting these approaches for any other desired switching voltage pattern. Using both analysis and simulations, PWM strategies will be described for both dc-dc and dc-ac applications using the full-bridge converter.

4.2   Topology and Conduction States

In this section, we will present the full-bridge converter. The full-bridge converter has vast applications in many domains of power electronics, may it be battery chargers, electric vehicles, renewable energy integration and many more. The converter is ideal when the intention is to interface a single dc voltage source either to a dc system or an ac system. Due to the popularity of this converter, it can be readily purchased as a module from many manufacturers of semiconductor components. For a newcomer to power electronics, becoming familiar with this converter is almost a rite of passage to becoming a true power electronics engineer.

Figure 4.1 shows the topology of the full-bridge converter. It is immediately evident that the full-bridge converter is merely an extension of the half-bridge converter in the previous chapter. It contains two converter legs instead of just one and the dc voltage source can be used directly across the terminals of the converter legs, though in many cases a dc capacitor is recommended to maintain a stable dc bus voltage. Since there are two converter legs, the ac circuit can be completed without any direct connection to the dc bus. This in turn minimizes the ripple in the dc bus voltage as the entire ac current is not flowing through the dc capacitor. The full-bridge converter is readily available as modules in various topologies, though the most popular are shown in Fig. 4.2—one where the terminals are already connected internally and the other where the terminals can be extended if needed [1–3].

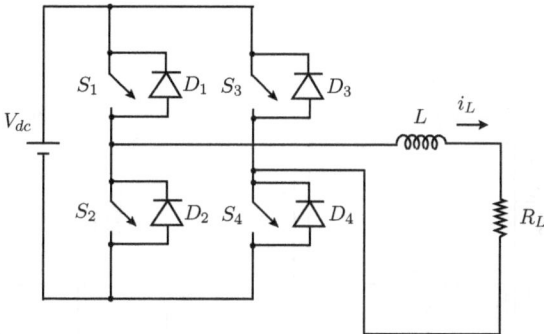

Fig. 4.1 Full-bridge dc-ac converter topology

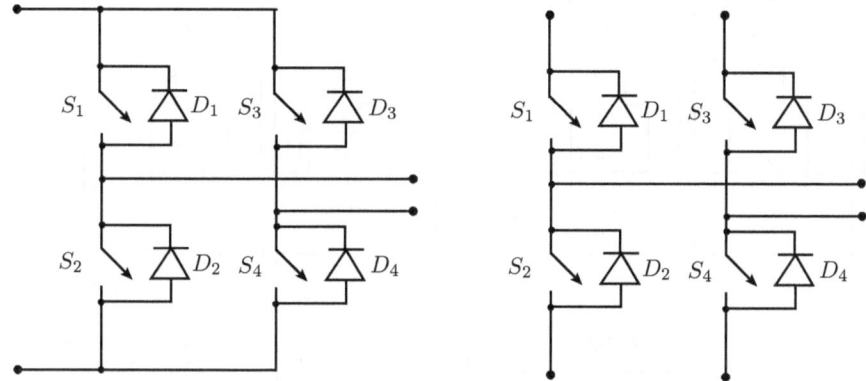

Fig. 4.2 Full-bridge dc-ac converter modules

Figure 4.3 shows some of the conduction states of the full-bridge converter. On the left are the conduction states during the positive half-cycle of the inductor current, when the inductor current i_L is positive in the direction shown, while on the right are the conduction states during the negative half-cycle of the inductor current, when the inductor current is negative in the direction shown. In the upper left, the controllable devices S_2 and S_3 are turned OFF but the devices S_1 and S_4 are turned ON. The dc voltage source acts as a driving force, and the inductor current increases in magnitude. The lower left shows the conduction state when the controllable devices S_1 and S_4 are subsequently turned OFF. Due to the inductor current being positive in the direction shown, the controllable devices S_2 and S_3 cannot conduct even if they are turned ON, and therefore, the inductor current can only freewheel through the diodes D_2 and D_3. In this state, the dc voltage source is a retarding force, and as a result the inductor current i_L decreases in magnitude. The reader can verify that these conduction states are quite similar to the conduction states of the half-bridge converter in the previous chapter, except that the ac current flows through two power devices rather than through the dc bus.

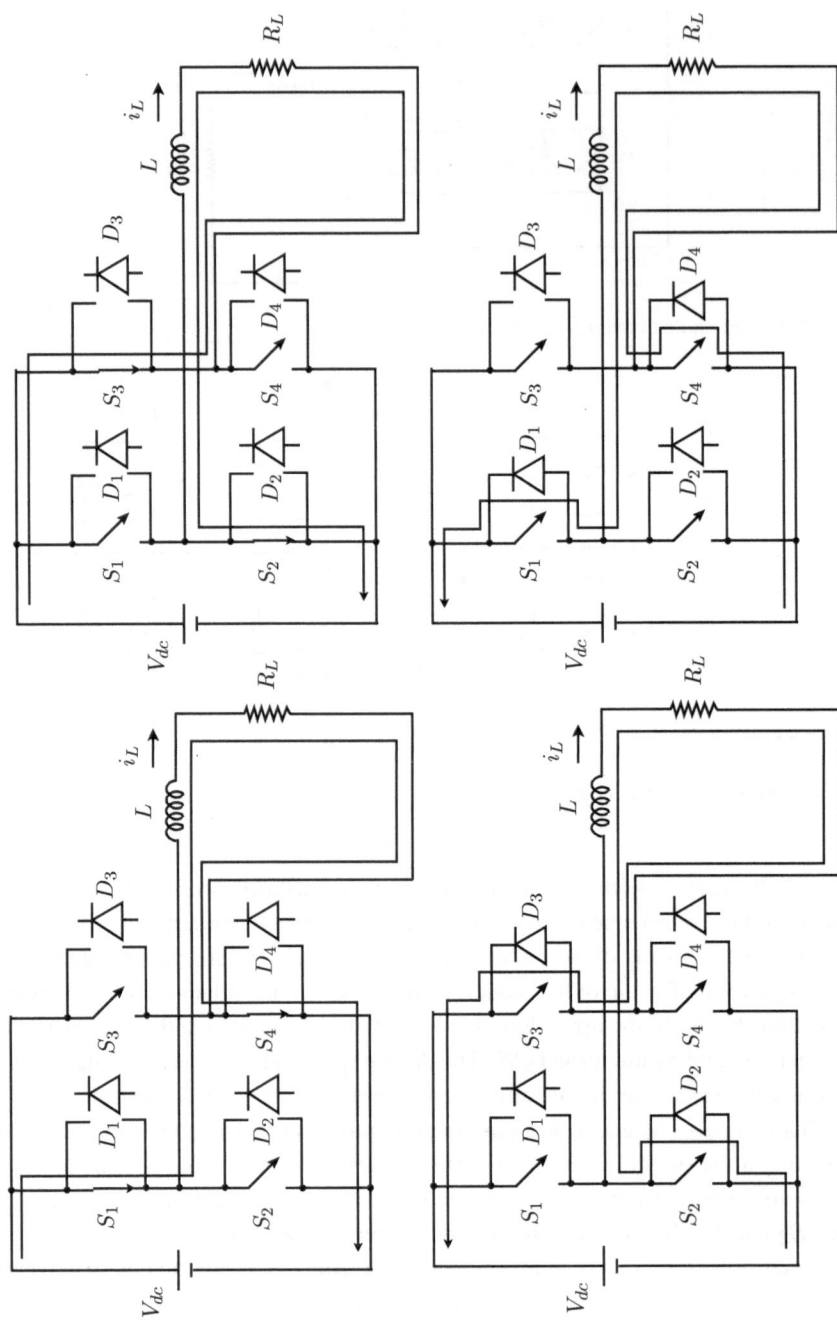

Fig. 4.3 Full-bridge dc-ac converter basic conduction states

The upper right of Fig. 4.3 shows the conduction state when controllable devices S_2 and S_3 are turned ON while controllable devices S_1 and S_4 are turned OFF. As this is during the negative half-cycle of the inductor current i_L, the dc voltage acts as a driving force, and causes the inductor current to increase in magnitude though in the negative sense. The lower right shows the conduction state when the controllable devices S_2 and S_3 are subsequently turned OFF. Since the inductor current is negative, the controllable devices S_1 and S_4 cannot conduct even if they are turned ON, and therefore, the inductor current i_L has to freewheel through the diodes D_1 and D_4. As can be seen, the dc voltage source acts as a retarding force, causing the inductor current to decrease in magnitude. Therefore, similar to the positive half-cycle, the conduction of the controllable devices causes the inductor current to increase in magnitude, while the inductor current will decrease in magnitude when the current freewheels through diodes.

Figure 4.3 is merely a replica of the conduction states of the half bridge converter shown in the previous chapter. However, when one takes into account that the full-bridge converter has twice the number of controllable devices and associated anti-parallel diodes, one needs to wonder if there are other conduction states due to the different conduction combinations possible. As always, two controllable devices in a single leg cannot conduct simultaneously to prevent a short-circuit of the dc voltage source, and therefore, that combination continues to be a forbidden state for the full-bridge converter as well. But, instead of only the diagonal devices conducting simultaneously, it is also possible for only the upper devices or the lower devices to conduct. Figure 4.4 shows these conduction states. In these conduction states, one will find a controllable device conducting simultaneously with an anti-parallel diode. The reader is once again advised to draw these conduction states separately with a pencil and paper to understand these additional states.

In the left of Fig. 4.4, the controllable devices S_1 and S_3 are intermittently switched while devices S_2 and S_4 are turned OFF—which implies only the upper devices of the full-bridge can be conducting. In the right of Fig. 4.4, the controllable devices S_2 and S_4 are intermittently switched while the devices S_1 and S_3 are turned OFF—which implies only the lower devices of the full-bridge can be conducting. This is perfectly safe, as the dc voltage source is not short-circuited. These conduction states do not feature the dc voltage source at all, and therefore, in these states, it is the inductor current that alone is freewheeling. Though the upper devices or the lower devices are turned ON simultaneously, only one of them can conduct depending on the direction of the inductor current, while the anti-parallel diode across the other controllable device will conduct. One will find some switching strategies that use these additional states, while some other switching strategies do not use these at all. The significance of these conduction states will be described fully in the next section.

The simulation of the basic conduction states of Fig. 4.3 can be found in the folder `full_bridge_dc_ac_bipolar_pwm` within the folder `chapter4` at the link: https://github.com/opensourceelectrical/switching-strategies-for-power-electronics.

The additional states in Fig. 4.4 will not be found in this simulation and will be deferred until the next section. In order to regulate the conduction states of the controllable devices,

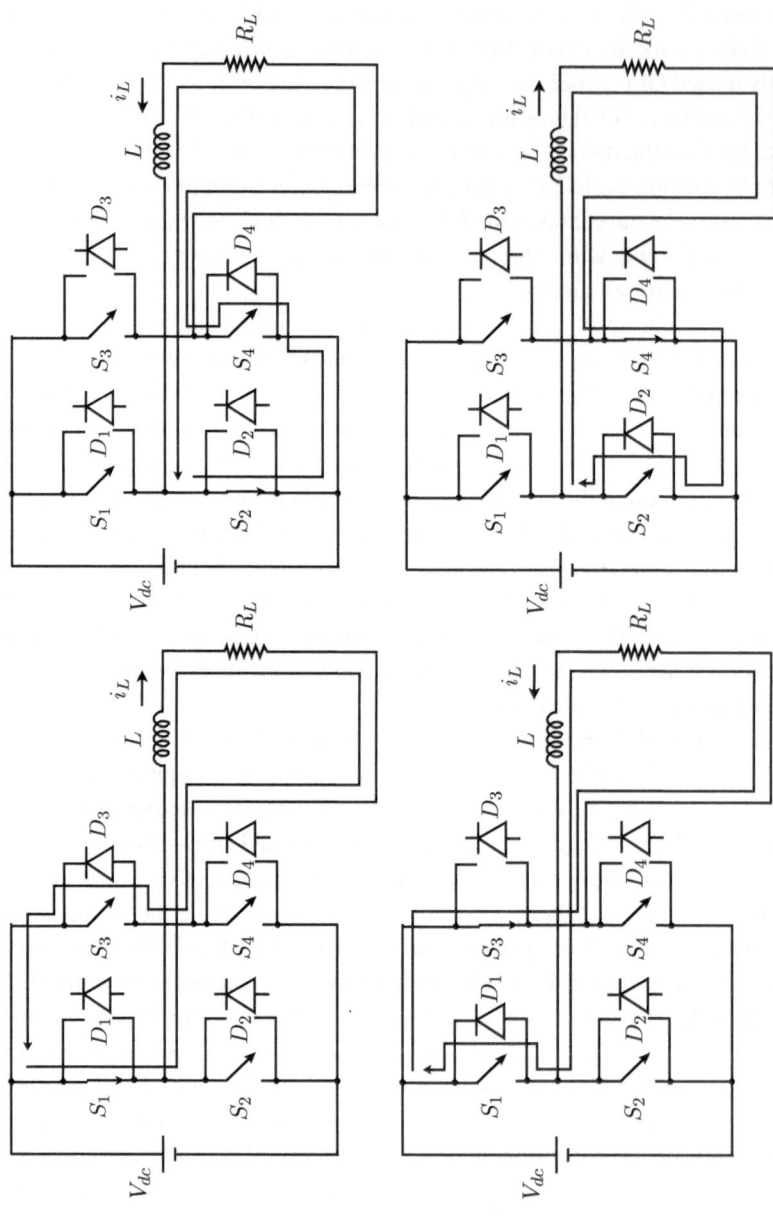

Fig. 4.4 Full-bridge dc-ac converter additional conduction states

Fig. 4.5 Full-bridge converter output current

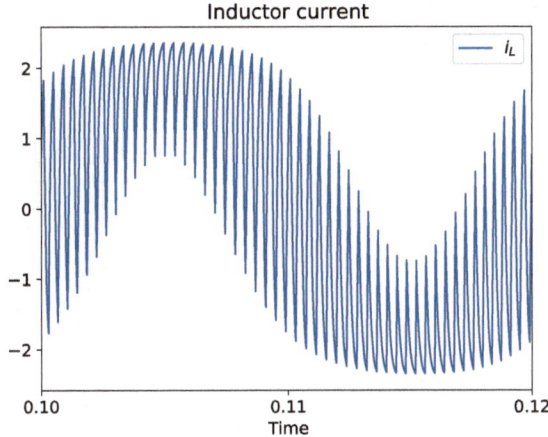

Fig. 4.6 Full-bridge converter output voltage

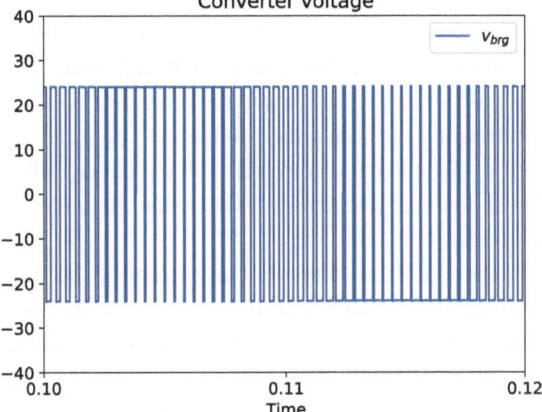

sine-triangle PWM will be used similar to the half-bridge converter of the previous chapter. For a modulation signal $m(t)$ and a carrier signal $c(t)$, the switching logic is as follows:

When $m(t) > c(t)$, S_1, S_4 are ON, S_2, S_3 are OFF

When $m(t) < c(t)$, S_1, S_4 are OFF, S_2, S_3 are ON.

The reader is advised to confirm that the above switching logic will result in the conduction states of Fig. 4.3. As with the case of the half bridge converter in the previous chapter, the full-bridge converter merely supplies a resistor-inductor load, as connecting the converter to a single-phase grid will need a phase-locked loop and closed-loop control.

Figure 4.5 shows the inductor current over a cycle. The switching frequency of the converter has been decreased to 2500 Hz for a clearer plot of the inductor current over a cycle. Figure 4.6 shows the converter output voltage over the same cycle. The reader can also attempt to merge the two plots to examine how the converter voltage impacts the change in inductor current. Figure 4.6 shows that the converter output voltage contains only two

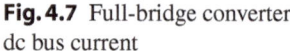

Fig. 4.7 Full-bridge converter dc bus current

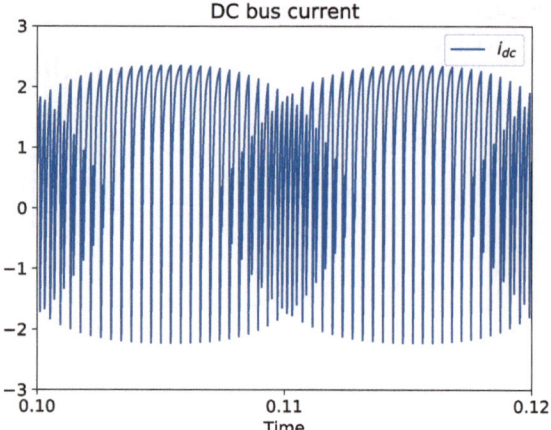

values, which are $+V_{dc}$ and $-V_{dc}$. The dc voltage source appears in the conduction loop as either a driving force or a retarding force. The reader is encouraged to verify that in the positive half-cycle, when the converter voltage is $+V_{dc}$, the inductor current increases, while when the converter voltage is $-V_{dc}$, the inductor current decreases. In the negative half-cycle, when the converter voltage is $-V_{dc}$, the inductor current increases in magnitude but in the negative sense, while when the converter voltage is $+V_{dc}$, the inductor current decreases in magnitude.

Figure 4.7 shows the dc bus current of the full-bridge converter. In contrast to the half bridge converter in which the ac current completes its path through the dc bus capacitors, in the full-bridge converter, the current flowing in the dc bus is primarily a ripple current. However, to fully understand the nature of the current, the reader is encouraged to zoom into the plot of the dc bus current and compare it with the converter voltage. At first glance, the dc bus current appears to be similar to an amplitude modulated waveform. However, this is merely due to the variation in the ripple in the inductor current. The ripple will be of large magnitude when the pulse width of the converter voltage is large and this typically occurs when the modulation signal is close to +1 or −1. Moreover, it may appear that the dc bus current may have an average zero value. However, upon close inspection, the reader will find that the positive peak of the dc bus current is in fact larger than the negative peak of the dc bus current.

The question then arises—what determines the average value of the dc bus current? For this, we must apply the law of energy conservation. The converter is primarily supplying the resistor load which for now we have assumed to be merely an ohmic loss. In addition, the converter will supply the losses in the system—losses in the inductor, losses in the converter devices and losses in the dc link. Therefore:

$$\text{Power supplied} = V_{dc} I_{dc}^{av} = I_L^2 R_L + \text{Losses} \qquad (4.1)$$

With the above expression, the average value of the dc bus current can be determined. If the converter were interfaced to the grid, it would also be possible for the converter to receive energy from the grid, in which case, the above equation will need to be rewritten accordingly.

In this section, we examined the basic operation of a full-bridge converter supplying a resistor-inductor load. The reader will find many aspects of the converter operation similar to that of the half bridge converter in the previous chapter. The simulation presented in this section will show the primary difference in the operation of the full-bridge converter and that of the half-bridge converter to be the dc bus current. In the next section, we will begin with our representation of the converter output voltage as a vector, and describe the significance of the additional conduction states that were not simulated in this section. This in turn will lay the foundation of the analysis that will follow in the later chapters.

4.3 Vector Representation of the Converter Output Voltage

In the previous section, a very basic simulation of the full-bridge converter was presented. In this simulation, the converter output voltage was either $+V_{dc}$ or $-V_{dc}$. This form of converter switching is often called as bipolar PWM, as the pulses have usually two extreme values. In this section, we will gradually introduce the impact of the additional conduction states of the full-bridge converter, through another PWM strategy called as unipolar PWM. The choice of PWM strategy is determined by the application of the converter. In certain applications, it will be imperative to utilize the additional conduction states, while in certain other applications, it might be beneficial in terms of decreasing the converter losses, and there might also be applications where these additional states cannot be used. Over the remainder of this chapter, a number of different examples will be presented for many of these scenarios.

The reader should review the simulation results of the previous section and compare the converter output of Fig. 4.6 with the conduction states of Fig. 4.3. As already stated in the previous section, for the additional conduction states of Fig. 4.4, the dc voltage source does not appear in the current loop. Therefore, for these additional conduction states of Fig. 4.4, the converter output voltage will be 0. This will soon be demonstrated through a simulation. However, if we were to list the possible output voltages that can be produced by the converter, these are now $+V_{dc}$, 0 and $-V_{dc}$. These three voltage levels can be represented as three vectors as shown in Fig. 4.8 along with the device combinations that result in a particular voltage level. The zero output voltage of the converter is represented by a zero vector, while the other two vectors have been represented by vectors of magnitude V_{dc} but in opposite directions.

It should be noted that such a vector representation was also possible for the bipolar PWM and also for the dc-dc converters presented before. Either the converter voltage would have two states $+V_{dc}$ and $-V_{dc}$ as with the case of the full-bridge converter with bipolar PWM or the half bridge converter in the previous chapter, or just $+V_{dc}$ and 0 as with the case of

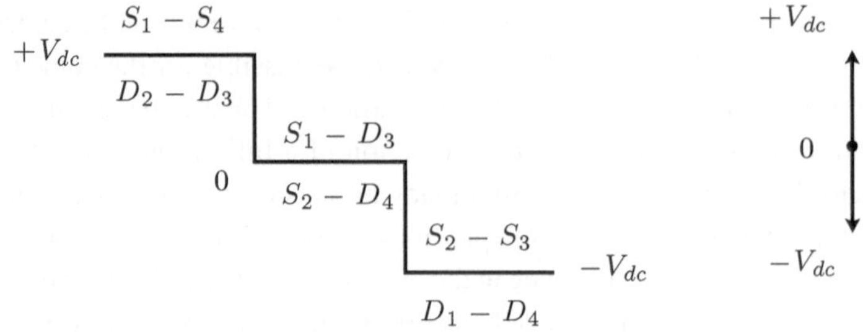

Fig. 4.8 Voltage vectors and voltage waveforms

the dc-dc converters presented in the previous chapter. However, as will be shown soon, this vector representation begins to truly be useful only when there are three or more states in the converter output voltage. This vector representation will also serve as a foundation for analysing the operation of converters in the later chapters which will have several possible output voltages.

Let us now use these voltage vectors to synthesize switching patterns. Figure 4.9 shows a few vector transitions representing device switchings for the full-bridge converter. On the extreme left of Fig. 4.9 is shown the bipolar PWM of the previous section. Since in a bipolar PWM, the converter output voltage changes between $+V_{dc}$ and $-V_{dc}$, the converter output changes between the $+V_{dc}$ vector and the $-V_{dc}$ vector. The transition has been shown only for the positive half-cycle when the inductor current is positive. Therefore, as can be seen from the converter output voltage waveform below the vector diagram, the inductor current flows either through the controllable devices S_1 and S_4 or through the diodes D_2 and D_3. We could introduce the zero vector into the switching process in a number of ways depending on the final voltage that we would like to achieve.

In the centre of Fig. 4.9 is shown one possibility of introducing a zero vector. During the positive half-cycle of inductor current, the voltage vector begins the switching cycle with the zero vector, changes to the $+V_{dc}$ vector and towards the end of the switching cycle changes back to the zero vector. Therefore, the voltage waveform remains symmetrical across the switching cycle as with the case of bipolor PWM. From the output voltage waveforms below the voltage vectors, as the switching occurs during the positive half-cycle, the inductor current flows either through the controllable device S_1 and diode D_3 or through the controllable devices S_1 and S_4. Since the inductor current does not face a retarding voltage $-V_{dc}$ during the zero voltage, the decrease in the inductor current will be significantly less than the decrease during case of the bipolar PWM. Moreover, in some applications such as dc-dc converters, it might not be necessary to produce a negative output voltage, and therefore, such a switching strategy might be more useful than bipolar PWM.

Besides the change in output voltage, unipolar PWM shown in the center of Fig. 4.9 presents another advantage of lower switching losses. When the inductor current that was

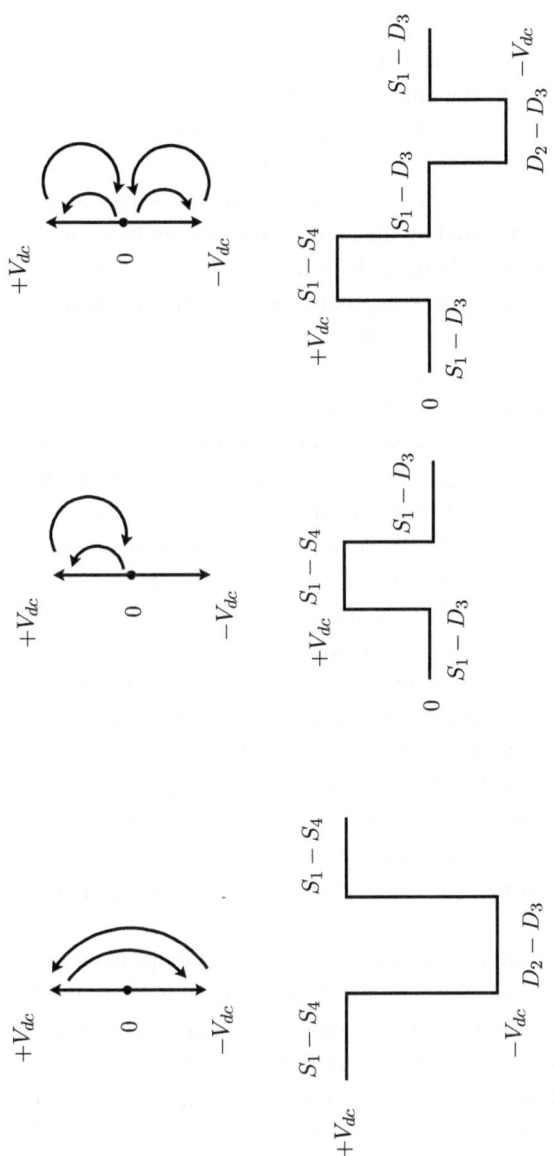

Fig. 4.9 Voltage vector transitions and corresponding waveforms

flowing through controllable devices S_1 and S_4 begins to flow through devices S_1 and D_3, the controllable device S_4 has turned OFF and the diode D_3 has started conducting. Therefore, device S_4 and diode D_3 experience switching losses. In comparison to bipolar PWM shown in the left of Fig. 4.9, the controllable devices S_1 and S_4 turn OFF and diodes D_2 and D_3 start conducting. Therefore, four devices experience switching losses in bipolar PWM instead of just two in unipolar PWM. Unipolar PWM can therefore, not only reduce the ripple in the current due to the nature of the converter output voltage, but can also decrease switching losses and subsequently increase the efficiency of the converter.

Another popular way of achieving PWM using the zero vector is as shown in the right of Fig. 4.9. In this case, both the $+V_{dc}$ vector and the $-V_{dc}$ vector can be found in a single switching cycle separated by the zero vector. As can be seen, the voltage waveform is symmetrical about the switching cycle as long as the time duration of the $+V_{dc}$ pulse is equal to the time duration of the $-V_{dc}$ pulse. This switching strategy is popular in many high-frequency resonant converters that use transformers either for isolation or for voltage transformation, as the average value of the output voltage is zero over a switching cycle, and therefore, prevents saturation of the transformer. This form of PWM is also called phase-shift PWM and will be demonstrated through a simulation in a later section in this chapter.

A quick examination of Fig. 4.9 and the description above might seem like we have unnecessarily complicated the process of converter switching. However, the greatest advantage of using such a vector representation of the converter voltage is to produce a simplified analytical method to generate the switching states necessary during any given switching cycle. To be able to describe this advantage, we would need to introduce a few details of control which we have avoided so far. In any application that uses a power electronic converter, closed loop control is essential to ensure that either the final output voltage or the final output current or any other variable of interest is equal to a particular desired value. As an example, if the converter is used as a power supply, the output voltage must be closely regulated to a desired value despite any changes in the load connected to the power supply. If the converter is used for grid integration of a solar PV panel, the current injected into the grid must meet grid quality standards - must be of grid frequency, must have a harmonic content lower than a level established by regulations, and also other requirements as established by the grid utility.

Closed-loop control will require one or more control loops in order to be able to regulate the desired variable. The number of control loops is usually determined by how the different variables in the system are related to each other. The control input or the control action is the converter voltage or the switching strategy that produces this converter voltage, while the final output can be the variable that is desired to be regulated such as the output voltage, the grid current or even the motor speed in the case of an electric vehicle. Therefore, in the event that the final variable that needs to be regulated is indirectly related to the converter output voltage, there can be many control loops that are needed to achieve the regulation. However, without going into the details of control loop design, since the primary control action is the converter voltage, the closed-loop control will produce in many cases, the required converter voltage that must be generated.

In most modern applications, closed-loop control is implemented digitally using micro-controllers or equivalent platforms such as Field Programmable Gate Array (FPGA) controllers. The control loop requires different system variables to be measured at regular intervals. This interval can be either equal to the switching time period, or can be a multiple of the switching time period. One would expect at the end of every control time interval, a new value for the control action which will remain constant for the duration of the next control time interval. Therefore, in power converter control, a new value of the required converter voltage is available once in every control time period. Depending on whether this control time period is equal to or greater than the switching time period, there will be a new converter voltage available either every switching cycle or once every few switching cycles. However, if we get back to our focus on converter switching and not closed-loop control, it is safe to state that in every switching cycle, one can expect a value of the converter voltage that needs to be generated through an appropriate switching logic.

Now that we have a value of the converter voltage that we would like to achieve in a particular switching cycle, the next question is, on how we will do so. The power converter can only produce voltage levels such that the time average of all the levels in a cycle is equal to the desired voltage required to be produced. If we denote the required converter voltage to be V_{req}, and the switching time period to be T_{sw}, we can think of the following general equation for the power converter:

$$\sum V_k T_k = V_{req} T_{sw} \tag{4.2}$$

Therefore, we can think of various combinations of voltage vectors for certain time intervals such that the sum of the product of the vectors and the time interval for which they are produced is equal to the product of the required converter voltage and the switching time period.

For the full-bridge converter that has been described in this section, the general equation can be written as:

$$(+V_{dc})T_+ + (0)T_0 + (-V_{dc})T_- = V_{req} T_{sw} \tag{4.3}$$

From Fig. 4.9, the values of T_+, T_0 and T_- will change for each of the three modulation strategies shown. For the bipolar PWM, T_0 will be zero as the zero vector is never produced. For the unipolar PWM in the center of Fig. 4.9, either T_+ or T_- will be zero, as only one non-zero voltage vector is produced in combination with a zero vector. Moreover, the time interval T_0 of the zero vector is divided into two separate time intervals of $\frac{T_0}{2}$. For the phase-shift control in the right of Fig. 4.9, the time interval T_+ of the $+V_{dc}$ vector is equal to the time interval T_- of the $-V_{dc}$ vector, while the time interval T_0 of the zero vector is divided into three separate intervals. In this case of the phase-shift control, the equivalent output voltage over a switching cycle will be zero, and therefore, it becomes necessary to adapt the above equation to only half of the switching time period.

Once the voltage vectors and their time intervals are determined, it is possible to implement the switching necessary to produce these vectors through a digital microcontroller. Many modern microcontrollers specifically intended for power electronic applications have specialized modules that make the implementation extremely simple. For the case of a full-bridge converter, it might appear to be an overkill to use such an approach to generate the switching logic. However, in the more advanced converter topologies that will follow in the later chapters, it will become evident that using such an approach greatly simplifies converter switching and allows us to produce switching logic in a systematic and repeatable manner. In the case of multiple switching combinations that can produce a particular voltage vector such as for example the case of the zero voltage vector, we can choose the switching logic that results in the minimum number of device transitions to reduce converter losses.

In this section we introduced the concept of voltage vectors to represent the converter output voltage. With the case of a full-bridge converter that can produce three possible output voltages, such a vector representation allows us to synthesize different switching patterns while still producing the desired converter voltage as a time average over the switching cycle. This vector representation of the converter voltage is the basis for advanced switching strategies such as Space Vector Modulation (SVM) that is used for multi-phase and multi-level converters as will be described in the later chapters. The remaining sections in this chapter will present simulations of various possibilities of converter switching using the full-bridge converter which can then be used as a foundation for the later chapters.

4.4   PWM Strategies for dc-dc Converters

In the previous section, we had described how the output voltage of the full-bridge converter can be expressed as a vector. Due to the fact that a full-bridge converter contains four controllable devices and four anti-parallel diodes, the number of possible conduction states is greater than the half-bridge converter examined in the previous chapter. Specifically, we found that in addition to the $+V_{dc}$ and $-V_{dc}$ output voltages that a half-bridge converter can generate, a full-bridge converter can generate a zero voltage as well. Using vectors diagrams, we examined potential transitions between the vectors, and created a mapping to output voltage patterns. We had introduced the concepts of unipolar PWM and phase-shift PWM, and some of the benefits of these kinds of strategies. In this section, we will examine the concept of unipolar PWM in detail in the context of dc-dc converters.

For the purpose of analysis and simulation, let us extend the simulation from the previous chapter of bidirectional buck-boost converter, except that in this section, we will consider two full-bridge converters that are interfaced through a single inductor. This is shown in Fig. 4.10. Since we have two full-bridge converters connected together, we can have bi-directional transfer of power and also complete control over the voltages produced by each converter. The reader should be familiar with the simulation of a bidirectional buck-boost converter comprised of two half-bridge converters presented in the previous chapter, such

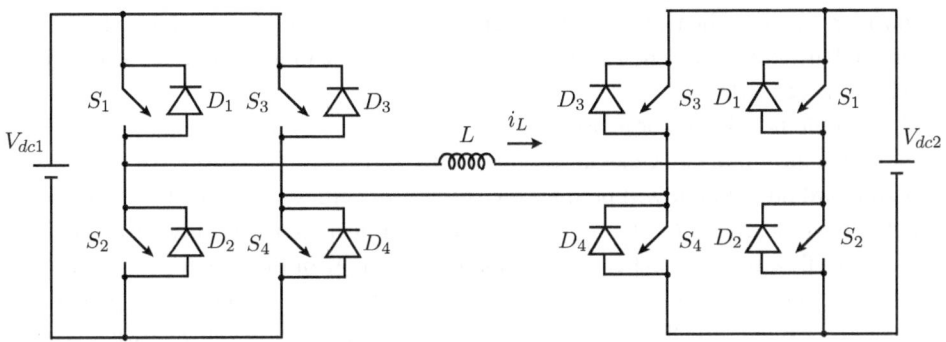

Fig. 4.10 Full-bridge converter interconnection

that the limitation of that system can be compared with the system presented in this section. For the sake of simplicity, let us assume that the dc voltage sources that supply the dc buses of both converter have the same level, say 24 V. Though in a practical system, one would expect two energy sources at different voltages to be interfaced through full-bridge converters. Before we present a simulation study with simulation results, let us examine all the different operating scenarios that are possible with Fig. 4.10.

Let us denote the full-bridge converter in the left of Fig. 4.10 to be the Left Converter (LC) and the full-bridge converter in the right of Fig. 4.10 to be the Right Converter (RC). Since we have two dc-dc converters, this is a perfect case for implementing unipolar PWM for switching the converters, as in order to achieve any dc output voltage, a converter need only produce two voltage levels—one of them being $0V$ and the other being $+V_{dc}$ or $-V_{dc}$. From the perspective of power flow, power can either flow from the LC to the RC, or from the RC to the LC. From the perspective of voltage, however, there are a number of cases, though, at first glance, it might not appear to be so. The simplest case being that both LC and RC produce output voltages that have positive averages. If the average output voltage produced by LC is greater than the average output voltage produced by RC, an average current will flow from LC to RC, and therefore, LC will supply power to RC. On the other hand, if the average output voltage produced by LC is lesser than the average output voltage produced by RC, an average current will flow from RC to LC, and therefore, RC will supply power to LC.

A more interesting case occurs when one of the converters produces an output voltage that has an average negative value. In such a case, there are two possibilities. The first, the other converter will also produce an output voltage with a negative average, such that the current flowing between the two converters is of a magnitude that would be considered normal operation. The second, the other converter produces an output voltage with a significant positive average value which results in a very large current flowing between the two converters. Though to fully describe such circumstances, one would need to describe the application in detail, a few examples will help to give the reader a clearer perspective. Let us consider the case when a dc battery is supplying a dc motor. Let us suppose both the dc battery and the motor have their own full-bridge converters, though, it might be possible to have just one converter.

If we would like the motor to accelerate, the dc battery will need to supply power to the motor so that the motor can generate an accelerating torque. If the motor were to be rotating in the forward direction, it will produce a positive emf across its terminals. Therefore, the output voltage produced by the motor converter will also have a positive average. Since current must flow from the battery to the motor, the average of the output voltage produced by the battery converter must be greater than the average of the output voltage produced by the motor converter. If on the other hand, we would like the motor to slow down, the output voltage of the battery converter can be adjusted to produce an average output voltage that would be lower than the average output voltage of the motor converter. Since the current would then flow from the motor to the battery, the negative torque would slow down the motor.

Let us now consider the extreme cases. If the motor were to be rotating in the forward direction, and we wish to suddenly stop it—think of an emergency brake on an electric vehicle. It would be necessary not only to produce a negative torque, but a very large negative torque. In such a case, the battery converter can be controlled to produce an output voltage with a large negative average, as a result of which, a large current will flow from the motor to the battery causing regenerative braking. In such cases, it will also be necessary to ensure that the current remains within safe limits and does not damage either the converters or the battery. The other extreme case would be when the motor were to be running in the reverse direction, thereby producing a negative emf across its terminals. If an emergency brake were to be applied in this case, a large positive torque would need to be produced, due to which the battery converter will need to produce an output voltage with a large positive average. In the case of a four-quadrant drive where both torque and speed can be negative, the converters will need to produce both positive and negative average output voltages while also allowing bidirectional flow of current.

With the above background, in the discussion to follow, we will use the operating condition to depict the output voltages that must be produced by the two converters LC and RC. Subsequently, we will infer the devices that will conduct in the two converters, and thereby produce the vectors needed to achieve the necessary output voltage. However, even though the vectors can be used directly along with precise timer intervals in a hardware implementation, the actual use of vectors will be deferred to the next chapter. In this chapter, we will examine how it is still possible to compare carrier waveforms with duty ratios to achieve the necessary output voltages. The reason for doing so is to show how it is possible to use various modules readily available in many microcontrollers in a fairly flexible manner to produce different switching pulse patterns. All that one needs to do is to determine how devices are to be turned ON and OFF, and accordingly formulate a switching strategy for the purpose. Therefore, this chapter will conclude with the flexibility of the PWM concept which is the reason for its popularity in power electronics.

Fig. 4.11 Voltage vectors for positive average voltages

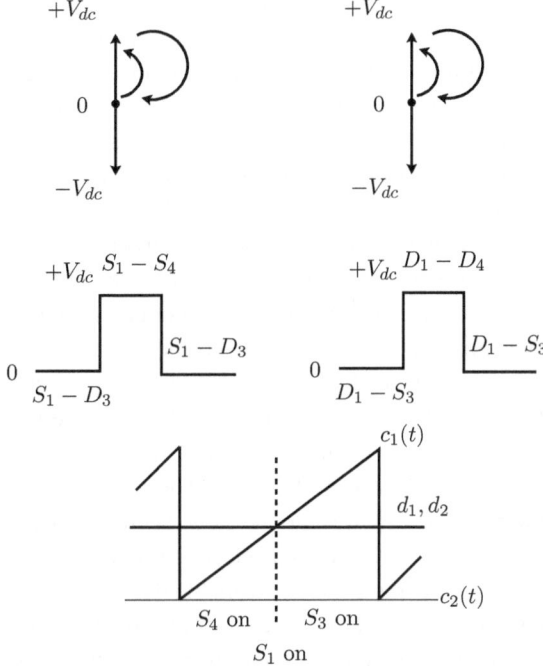

Let us consider the first case where both LC and RC are producing output voltages with positive averages, and the current is flowing from LC to RC. In such a case, the voltages produced by the converter are shown in Fig. 4.11. Both converters need only the $+V_{dc}$ vector and the zero vector. Since the output voltage of LC has a greater average than the output voltage produced by RC, the duty ratio of LC will be larger than RC, though this is not shown in Fig. 4.11. It is recommended for the reader to draw the circuit of Fig. 4.10 and map out the conduction paths. However, from the direction of current, one can quickly state that the devices $S_1 - S_4$ and $S_1 - D_3$ of LC will be conducting in the two intervals, and the devices $D_1 - D_4$ and $D_1 - S_3$ of RC will be conducting in the two intervals. The reader should note that even though the $+V_{dc}$ vector of RC is applied during an interval of the switching cycle, it is the anti-parallel diodes across the controllable devices which will conduct due to the direction of the inductor current. If the closed-loop control produces the duty ratio d_1 for LC and d_2 for RC, the switching signals can be generated for LC by choosing vector $+V_{dc}$ ($S_1 - S_4$) for d_1T and zero vector ($S_1 - S_3$) for $(1 - d_1)T$, and for RC by choosing vector $+V_{dc}$ ($S_1 - S_4$) for d_2T and zero vector ($S_1 - S_3$) for $(1 - d_2)T$.

Let us now formulate a strategy to generate the same switching pulses by comparing duty ratios and carrier waveforms. Since we need to generate a vector $+V_{dc}$ and a zero vector using devices S_1 and S_3, it easy to notice that the controllable device S_1 in both converters is always turned ON. Either the controllable device S_1 will conduct or the anti-parallel diode D_1 across it will conduct depending on the direction of current. Since device S_1 is always turned ON, the controllable device S_2 will be always turned OFF. The controllable devices S_3 and S_4 are turned on in a complementary manner, and for this, the logic can be the same as used for all the dc-dc converters before—when the duty ratio d_1 and d_2 are lesser than the sawtooth carrier waveform $c_1(t)$, the controllable device S_3 of converters LC and RC are turned ON, and when the duty ratios are greater than the sawtooth carrier waveform, the controllable device S_4 of converters LC and RC are turned ON. To ensure that the device S_1 of both converters are turned ON, we can compare the duty ratios d_1 and d_2 with a zero carrier signal $c_2(t)$. This is depicted in Fig. 4.11.

Let us now consider another case—when LC produces an output voltage with a negative average while RC continues to produce an output voltage with a positive average. The switching pulses of RC will be the same as in the previous case with a duty ratio d_2. However, LC will now be using the $-V_{dc}$ vector and the zero vector. This is shown in Fig. 4.12. The reader should note that because RC is producing an output voltage which a positive average while LC is producing an output voltage with a negative average, an average current will flow from RC to LC. Therefore, the devices conducting in RC are different from the previous case despite the same vectors being chosen, and this is due to the direction of current. The $-V_{dc}$ vector needs controllable devices S_2 and S_3 to be turned ON. For the zero vector, we will turn on controllable devices S_1 and S_3. To generate a zero vector, one could also choose to turn on the controllable devices S_2 and S_4. As before, it would be possible to generate the switching signals by turning ON the devices corresponding to the vectors for the corresponding time interval. However, let us formulate a strategy for generating switching pulses for LC which needs to produce an output voltage with a negative average.

In the previous case, the controllable device S_1 was continuously turned ON, and therefore, the duty ratio d_1 was compared with a zero carrier signal $c_2(t)$. However, in this case, controllable devices S_1 and S_2 are turned ON intermittently and of course, in a complimentary manner. Therefore, for this to occur, it is necessary to compare the duty ratio d_1 with a non-zero sawtooth carrier waveform. Since LC produces an output voltage with a negative average, the duty ratio d_1 produced by the closed-loop control will have a negative value. Therefore, the carrier signal $c_2(t)$ which was zero in the previous case is now updated to a negative sawtooth waveform as shown in Fig. 4.12. The switching pulses for the controllable devices S_1 and S_2 will now be generated by comparing the negative duty ratio d_1 and this negative sawtooth carrier waveform $c_2(t)$. The controllable devices S_3 and S_4 will switched as a result of the comparison between the duty ratio d_1 and the positive sawtooth waveform $c_1(t)$. Since the duty ratio which is negative, is compared with a sawtooth waveform $c_1(t)$ which is positive, the controllable device S_3 will always be turned ON. Therefore, we have formulated a switching strategy for the full-bridge converter in an iterative manner, such that it can function effectively in a four-quadrant operation mode using unipolar PWM.

With the above description of how we can formulate a unipolar switching strategy for a full-bridge dc-dc converter, let us now simulate the above two cases. The simulations can be found in the folder dual_dc_dc_conv_unipolar_pwm within the folder chapter4 at the link: https://github.com/opensourceelectrical/switching-strategies-for-power-electronics.

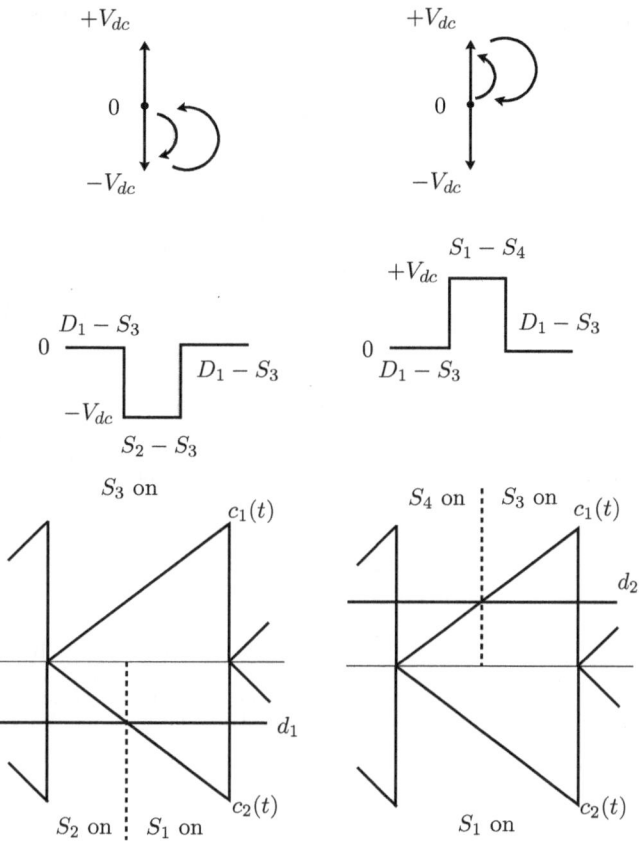

Fig. 4.12 Voltage vectors for positive and negative average voltages

The reader will also find another simulation case of a single full-bridge converter feeding a resistor-inductor load in the folder `single_dc_dc_conv_unipolar_pwm` within the folder `chapter4` at the link: https://github.com/opensourceelectrical/switching-strategies-for-power-electronics.

The reader is encouraged to examine that simulation in case the simulation of two interconnected full-bridge converters is too complicated to begin with. The reader should also note that this case of two full-bridge converters was specifically chosen to demonstrate the flexibility of the full-bridge converter in contrast to the half-bridge converter, even though a full-bridge converter contains twice the number of devices. In most applications that need a full-bridge converter, a single full-bridge converter is quite often sufficient, and unless necessary, one does not use two or more converters as in any system design, decreasing cost is always important.

Fig. 4.13 Converter output
voltages

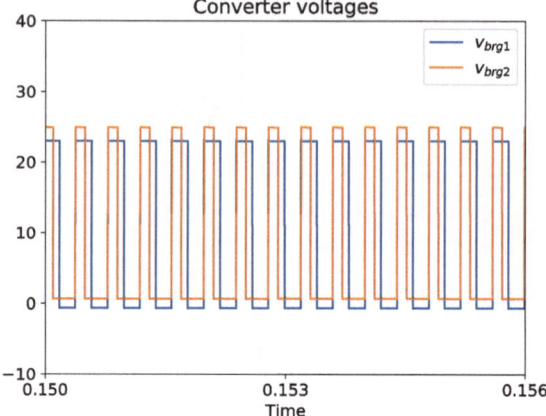

Fig. 4.14 Current flowing
between the converters

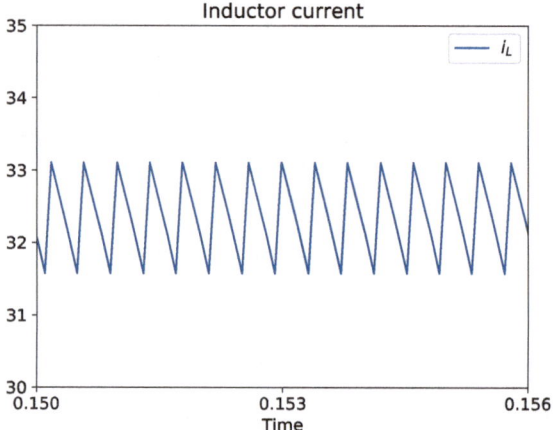

The reader will find two separate control files that generate the PWM signals for the two full-bridge converters. This is due to the fact that the two converters can be controlled in a fairly independent manner, which also includes phase-shifting the carrier waveforms or choosing a different switching frequency for the converters. Though this will not be shown in the simulation results, readers are encouraged to try these changes on their own. To begin with, let us examine the simple case of both converters producing output voltages with positive averages. For this purpose, let us set the duty ratio d_1 of LC to be 0.5 and the duty ratio d_2 of RC to be 0.3. The simulation results can be seen in Figs. 4.13, 4.14 and 4.15. Figure 4.13 shows the output voltages of the converters, where it can be seen that the $+V_{dc}$ pulse of LC is wider than the $+V_{dc}$ pulse of RC. Figure 4.14 shows the inductor current to have a positive average which implies that the inductor current is flowing from LC to RC.

Figure 4.15 shows the switching pulses generated for LC though RC will be quite similar as they are both generating an output voltage with positive averages. Only the switching

Fig. 4.15 PWM for LC

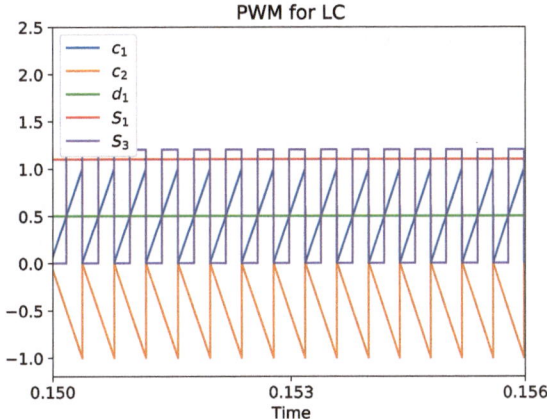

pulses for controllable devices S_1 and S_3 are shown, as the switching pulses for device S_2 will be complimentary to pulses for S_1, while the pulses for device S_4 will be complementary to pulses for device S_3. Due to the comparison of the duty ratio d_1 with the negative carrier waveform $c_2(t)$, device S_1 is continuously turned ON, while the comparison of d_1 with the positive waveform results in the device S_3 being intermittently turned ON. The reader should compare the output voltage v_{brg1} of LC with these switching waveforms to verify the results. As a closing note to this simulation result, the reader should note that the converter output voltages in Fig. 4.13 appear to have dc offsets. This is due to the fact that LC is supplying a current to RC, due to which the output voltages which are measured as floating voltages, exhibit ohmic drops due to the flow of this current.

Let us now examine the second mode of operation, where LC produces an output voltage with a negative average while RC will continue to produce an output voltage with a positive average. Figure 4.16 shows the converter output voltages, where it can be clearly seen that

Fig. 4.16 Converter output voltages

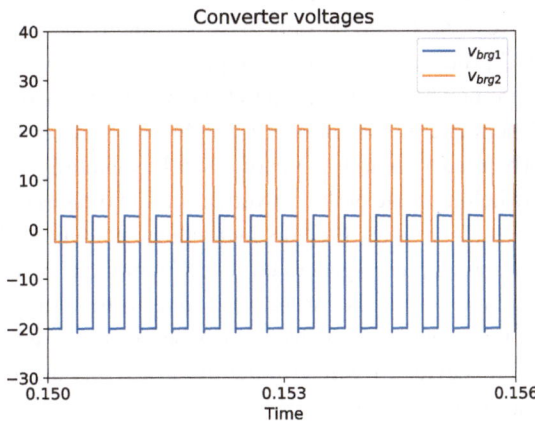

Fig. 4.17 Current flowing
between the converters

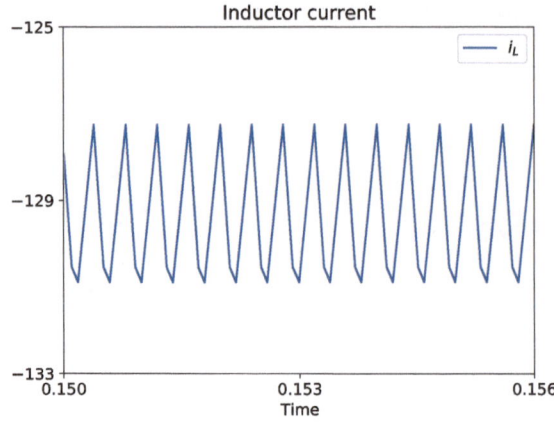

Fig. 4.18 PWM for LC

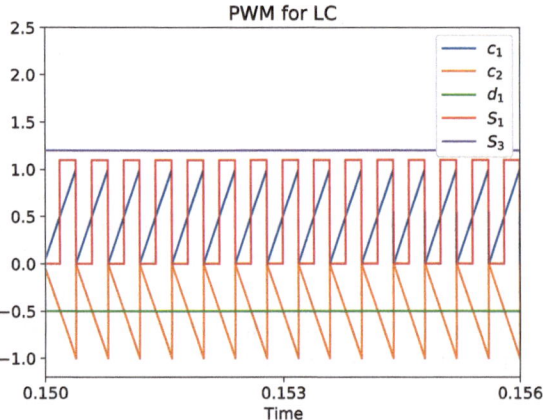

LC produces pulses of $-V_{dc}$ and the zero vector while RC continues to produce pulses of $+V_{dc}$ and the zero vector. Due to the current flowing from RC to LC, the output voltages have offsets which results in LC producing an output voltage with a positive dc offset, while RC produces an output voltage with a negative dc offset. Figure 4.17 shows the inductor current which can be seen to have a large negative average value, as it flows from RC to LC. The inductor current over a single switching cycle can be seen to have three separate $\frac{di}{dt}$s. The reader should compare a single switching cycle of the inductor current with a single switching cycle of the converter output voltages to verify that the voltage that appears across the inductor has three distinct values, namely $-2V_{dc}$, $-V_{dc}$ and 0. Figure 4.18 shows the switching pulses generated for LC while the switching pulses for RC will be similar to the previous case. As can be seen, due to the duty ratio d_1 being negative, the device S_3 is continuously turned ON as d_1 is being compared with the positive carrier waveform $c_1(t)$, while device S_1 is intermittently turned ON due to the comparison of d_1 with the negative carrier waveform $c_2(t)$.

This section described in detail the performance of power electronic converters with unipolar PWM. Since the application consisted of dc-dc converters, the choice of unipolar PWM was only natural. Using a sample two converter system, we demonstrated through analysis, how converters can be controlled to result in different operating conditions. For an operating condition, each converter needs a zero vector and either a $+V_{dc}$ vector or a $-V_{dc}$ vector. Once the duty ratio is obtained from a separate closed-loop controller, the switching pulses for the converters can be generated by turning ON the devices corresponding to a particular vector for a time interval proportional to the duty ratio. However, instead of choosing to directly apply the vectors, one can use carrier waveforms and compare the duty ratios with them in the way done for dc-dc converters before. One of the advantages of using carrier waveforms instead of vectors is to use readily available modules and algorithms when implementing unipolar PWM in hardware. Therefore, in this section, we described in a step-by-step manner, how it is possible to adjust carrier waveforms to achieve the same switching pulses as would have been obtained by using the voltage vectors. How the reader finally implements PWM in a hardware prototype is merely a question of convenience and using the peripherals of the available digital controller effectively.

4.5   PWM Strategies for dc-ac Converters

In the previous section, we had explored the concept of unipolar PWM in the context of full-bridge converters used in dc-dc applications. The focus on dc-dc applications enabled us to explore in detail how unipolar PWM can be used under different circumstances, as for dc-dc applications, unipolar PWM is sufficient for voltage regulation. However, in grid-frequency ac systems or pseudo ac systems (non grid-frequency or high frequency), a full-bridge converter will need to produce positive, negative and zero voltages under any given operating condition. Therefore, in the context of dc-ac applications, one needs to either modify unipolar PWM or use phase-shift PWM or bipolar PWM. In this section, we will initially explore the concept of unipolar PWM for full-bridge dc-ac converters in grid-frequency ac applications. We will go on to further discuss phase-shift PWM in detail in the context of dc-ac converters.

At the beginning of this chapter, we had provided a basic simulation for a full-bridge dc-ac converter using bipolar PWM. In the full-bridge converter of Fig. 4.1, the diagonally opposite controllable devices, S_1–S_4 and S_2–S_3 were turned ON and OFF together in pairs. The only possible output levels available from the full-bridge converter with bipolar PWM, thus, were $+V_{dc}$ and $-V_{dc}$. Though a bipolar PWM strategy appears to be the only natural option for switching a full-bridge converter supplying an ac system, a unipolar PWM strategy can result in several advantages as will be shown. In this section, we will extend the basic simulation for the full-bridge converter to include the conduction states from Fig. 4.4. In order to include all the possible conduction states, we will need to regulate the two legs of the full-bridge converter separately in a manner not very different from the previous section.

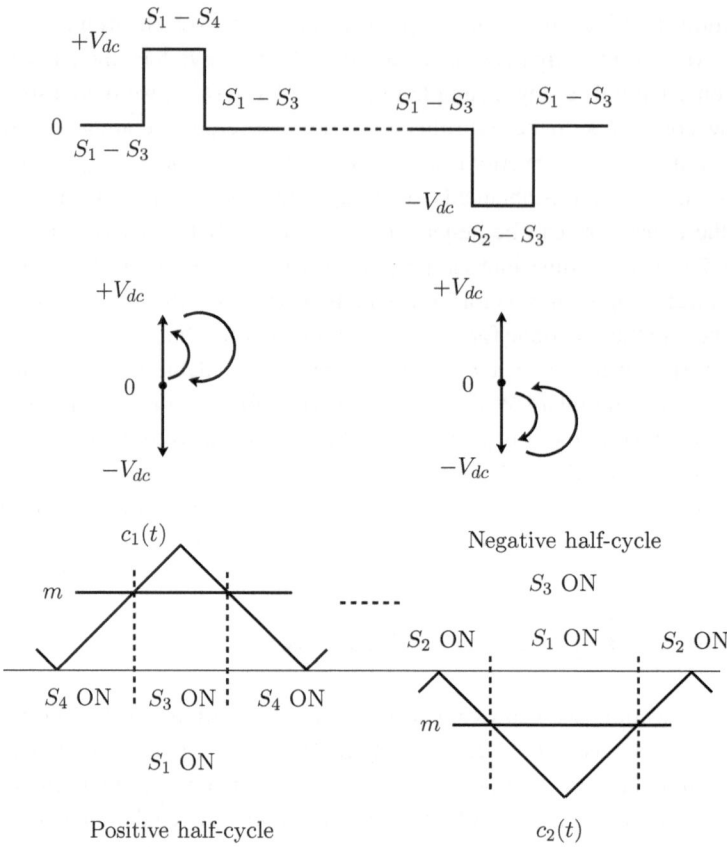

Fig. 4.19 Voltage vectors for full-bridge dc-ac converter with unipolar PWM

In a manner similar to the previous section, a unipolar modulation strategy can be formulated as a modification or extension of the bipolar PWM strategy. Figure 4.19 shows the unipolar PWM strategy where instead of using a single triangular carrier waveform, two triangular carrier waveforms $c_1(t)$ and $c_2(t)$ are used. One carrier waveform is from 0 to 1, while the other is from −1 to 0. Upon taking a closer look at Fig. 4.19, the reader will find that $c_2(t) = -c_1(t)$, in exactly the same manner as in the previous section. The switching rule can be listed as follows:

When $m(t) > c_2(t)$, S_1 is ON, S_2 is OFF
When $m(t) < c_2(t)$, S_1 is OFF, S_2 is ON
When $m(t) > c_1(t)$, S_3 is OFF, S_4 is ON
When $m(t) < c_1(t)$, S_3 is ON, S_4 is OFF

From the above switching strategy, it is quite obvious that when the modulation signal is positive, it will always be larger than the negative carrier signal $c_2(t)$. Therefore, the controllable device S_1 is continuously turned ON whenever the modulation signal is positive. Since the carrier waveform $c_1(t)$ is positive, the controllable devices S_3 and S_4 will be turned ON and OFF intermittently during the positive half-cycle of the modulation signal. During the positive half-cycle of the modulation signal, the $+V_{dc}$ vector is achieved with devices S_1–S_4 and the zero vector is achieved with devices S_1–S_3. During the negative half-cycle of the modulation signal, controllable devices S_1 and S_2 will be turned ON and OFF intermittently, while controllable device S_3 will always be turned ON. Therefore, the $-V_{dc}$ vector will be achieved through devices S_2–S_3 and the zero vector will be achieved by devices S_1–S_3. The reader must always keep in mind that even if a controllable device is turned ON, it will only conduct for a particular direction of the inductor current, and otherwise, the anti-parallel diode across the device will conduct instead.

The simulation of the full-bridge dc-ac converter using unipolar PWM can be found in the folder `full_bridge_dc_ac_unipolar_pwm` within the folder `chapter4` at the link: https://github.com/opensourceelectrical/switching-strategies-for-power-electronics.

The basic states in Fig. 4.3, as well as the additional states in Fig. 4.4 will both be found in this simulation. It is to be noted that when both the upper switches, S_1 and S_3, are ON, the output voltage is 0. Though the upper devices are turned ON simultaneously, only one of them can conduct depending on the direction of the inductor current, while the anti-parallel diode across the other controllable device will conduct. The output current circulates through S_1 and D_3, or D_1 and S_3 depending on the direction of the inductor current. A similiar condition occurs when both the bottom switches, S_2 and S_4 are on. These conduction states do not feature the dc voltage source at all, and therefore, in these states, it is the inductor current that alone is freewheeling.

Figure 4.20 shows the inductor current over a cycle. The switching frequency of the converter has again been decreased to 2500 Hz for a clearer plot of the inductor current over a cycle. In the case of unipolar PWM, when a switching occurs, the output changes

Fig. 4.20 Full-bridge converter output current

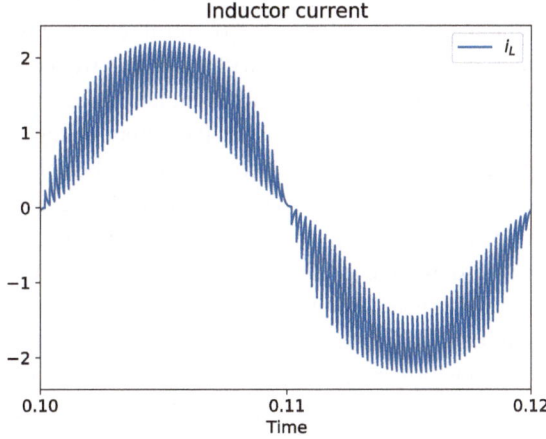

Fig. 4.21 Full-bridge converter
output voltage

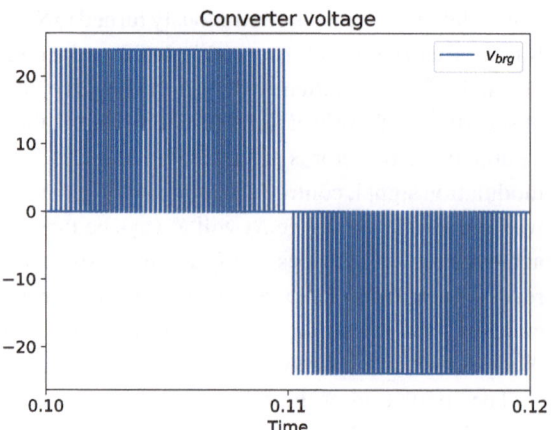

between 0 and $+V_{dc}$ or between 0 and $-V_{dc}$, thus, achieving three-level naturally sampled sine-PWM. Figure 4.21 shows that the converter output voltage now contains three values, which are 0, $+V_{dc}$ and $-V_{dc}$. The dc voltage source now only appears in the conduction loop as a driving force. The reader is encouraged to verify that in the positive half-cycle, when the converter voltage is $+V_{dc}$, the inductor current increases. In the negative half-cycle, when the converter voltage is $-V_{dc}$, the inductor current increases in magnitude but in the negative sense. The inductor current decreases in magnitude during either half-cycle as it freewheels when the converter voltage is 0.

In the context of unipolar PWM for ac systems, the above was an implementation for a full-bridge converter interfaced to a conventional grid-frequency ac system. The converter would therefore produce a switched ac voltage with a separate sequence for the positive half-cycle, and another sequence for the negative half-cycle. The PWM strategy will be fed with a modulation signal that has both positive and negative half-cycles, and all we needed to do was determine carrier waveforms and switching logic to be able to produce the switching combinations. The closed-loop control scheme will have a component that either determines the frequency of the grid or will generate a frequency in the case of grid-forming converters. The closed-loop scheme will eventually feed the PWM module with a modulation signal which has the same template as the output voltage that is desired. However, there is another application of the full-bridge converter where it is necessary to produce a voltage with a net zero average over a switching cycle. These applications are not in ac systems, but rather in dc systems, except that these systems use high-frequency transformers to either step-up or step-down the output voltage, or use transformers with multiple windings to produce multiple isolated output voltages.

In dc-dc applications that use transformers, the transformers are high-frequency ferrite core transformers as opposed to the laminated iron core transformers used in grid frequency applications. The frequency of operation of the dc-dc converters and the transformers can range from a few kilohertz to several megahertz. The final output or outputs of the dc-dc converter are dc voltages that are obtained by rectifying the voltages available at the windings of the transformers and filtering them with electrolytic capacitors. However, due to the use of a transformer, it is imperative that at no winding of the transformer should a voltage be applied that has a non-zero average voltage over a switching cycle, or else the transformer will get saturated. In the previous sections and chapters, dc-dc converters were controlled by regulating the duty ratio which determines the time interval for which a non-zero voltage ($+V_{dc}$ or $-V_{dc}$) is produced as an output. However, for the remaining time duration in the switching cycle, the output voltage is usually zero. Quite clearly, in such a case, the output voltage has a non-zero average over the switching cycle, and such a voltage cannot be applied to a transformer.

In order to produce a output voltage with a zero average over a switching cycle, it is essential that when the converter produces a non-zero voltage for any time interval, the converter also produces an opposite voltage for the same time interval. This has been shown in Fig. 4.9, where over a single switching cycle, the converter produces $+V_{dc}$ output and also $-V_{dc}$ output along with the zero vector. The reader should note that the time interval of $+V_{dc}$ must be equal to the time interval of $-V_{dc}$ for the average voltage over a switching cycle to be zero. It is also important to note that such a voltage pattern must be produced with the only control input being the duty ratio as before, as this duty ratio determines the time interval of one of the non-zero voltages. The only difference being that the duty ratio must be lesser than 0.5, as we must take into account that we need non-zero output voltages for twice the duration corresponding to the duty ratio. We must devise a PWM strategy which can produce the $+V_{dc}$ (or $-V_{dc}$) voltage for the desired time interval, as a well a $-V_{dc}$ (or $+V_{dc}$) voltage for the same time interval.

We can follow the strategy as before, and begin with the voltage pattern and subsequently decipher the switching combinations that produce the voltages. This is shown in Fig. 4.22. We must now solve the rest of the puzzle - given a duty ratio, how do we achieve this voltage pattern? One trick to understanding the puzzle is that when one wishes to produce symmetric voltage patterns such as these, in reality, each leg is being switched such that one controllable device is turned ON for half the time duration and the other controllable device is turned ON for the remaining half of the time duration. The deciding factor on the nature of the output voltage is the phase shift between the two legs of the full-bridge converter. This is shown in Fig. 4.22 below the device combinations. To achieve 50% duty ratio for each leg, the duty ratios d_1 and d_2 will have to be 0.5 in contrast to the previous cases. The modification will be made to the sawtooth carrier waveforms $c_1(t)$ and $c_2(t)$. In the previous examples, $c_2(t) = -c_1(t)$, and one carrier waveform is usually the mirror image of the other across the time axis. In phase-shift PWM, the duty ratio will be introduced as a phase shift in the carrier waveforms, with carrier waveform $c_2(t)$ being phase advanced by an amount equal to the duty ratio d. Once the carrier waveforms are phase shifted, carrier waveform $c_1(t)$ must be compared with 0.5, and carrier waveform $c_2(t)$ must be compared with -0.5 in a strategy similar to the previous cases:

When $-0.5 > c_2(t)$, S_1 is ON, S_2 is OFF

Fig. 4.22 Voltage vectors for
full-bridge dc-dc converter
with phase-shift PWM

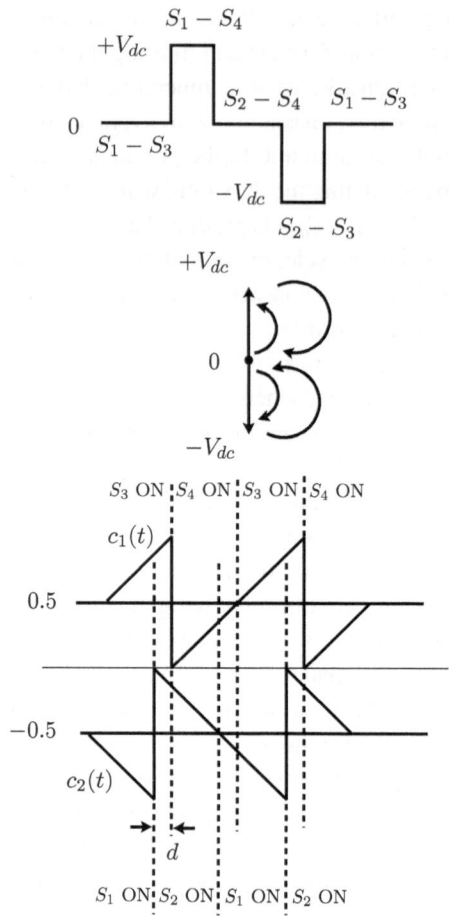

When $-0.5 < c_2(t)$, S_1 is OFF, S_2 is ON
When $+0.5 > c_1(t)$, S_3 is OFF, S_4 is ON
When $+0.5 < c_1(t)$, S_3 is ON, S_4 is OFF

The simulation for the phase-shift PWM can be found in the folder
`dc_dc_conv_phase_shift_pwm` within the folder `chapter4` at the link: https://
github.com/opensourceelectrical/switching-strategies-for-power-electronics.

Figures 4.23 and 4.24 show the simulation results. Figure 4.23 shows the PWM strategy
with the carrier waveforms being phase shifted by an amount equal to the duty ratio of 0.2.
Comparing these two carrier waveforms $c_1(t)$ and $c_2(t)$ with +0.5 and −0.5 respectively
produces phase shifted switching pulses. By controlling a leg of the converter as a basic
element, we ensure that two controllable devices in a leg cannot be turned ON simultaneously,
while also allowing for a certain degree of freedom in how the legs are switched with respect
to each other. Figure 4.24 shows the voltage produced by the full-bridge converter, and it

Fig. 4.23 Phase-shift PWM

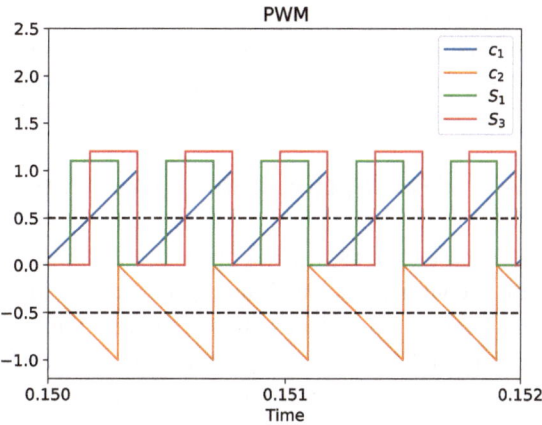

Fig. 4.24 Converter voltage

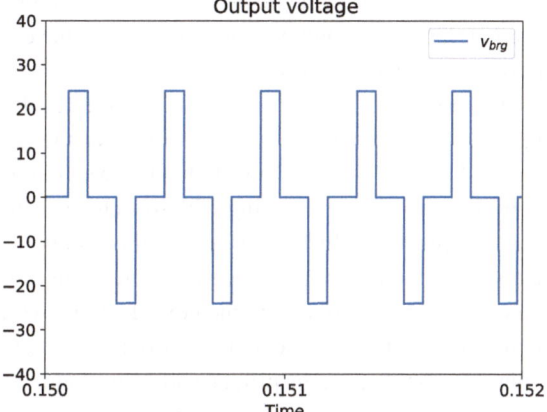

is very clear that in every switching cycle, there is one $+V_{dc}$ pulse and one $-V_{dc}$ pulse separated by zero vectors. The reader is encouraged to repeat the simulation varying the duty ratio and examining the effect on the carrier waveforms as well as the final output voltage produced by the converter. The simulation does not contain a transformer, as the magnetic modelling a transformer is a very different domain which will not be covered in this book.

In this section, we examined how modified PWM strategies can be formulated for a full-bridge converter functioning as a dc-ac converter or as a dc-dc converter but in an application that uses transformers. In both cases, we used the same approach to developing a PWM strategy—start with the desired output voltage waveform, map out the corresponding device combinations, and finally choose carrier waveforms and modulation signals/duty ratios whose comparison can result in the desired output voltage. In a manner similar to the previous section, we found the approach to formulating PWM strategies step-by-step and systematic, which leads us to conclude that this method would very flexible in various applications in power electronics. In the next section, we will summarize the different strategies examined so far and comment on the potential implications with respect to the ease or difficulty of hardware implementation using readily available microcontroller kits.

4.6   Closing Thoughts on the Strategies Discussed

In the past sections, we described several strategies to generate switching pulses for a full-bridge converter used in different applications. If one were to merely extend the half-bridge converter of the previous chapter, it would appear that bipolar PWM would be the most obvious and simplest approach to generate switching pulses for the full-bridge converter. However, as described in the previous chapter, greater number of devices implies a greater number of possibilities for switching combinations. Over several sections, we have gradually explored different strategies to arrive at the output voltage pattern that we desired. In this section, let us take another look at what we have achieved and the implication in terms of power converter operation.

   The essence of all the strategies examined in the previous sections was to able to use the three distinct output voltage possibilities of the full-bridge converter, namely $+V_{dc}$, $-V_{dc}$ and 0. In the second and third chapter, we found only two possible output voltages of a converter. It is only beginning with this chapter, that we examined the possibility of a third state of operation which occurs due to the conduction of controllable devices. This third state came into being due to the degree of freedom in operating the two legs of the full-bridge converter. As we progress to the next chapters, we will find several other states of operation that result in other distinct converter output voltages. However, before progressing to those converter topologies, the primary objective behind this chapter was to address the question - how can we devise switching strategies that allow us to capture the different possible output voltages of the converter?

   Before we address the question how, let us again return the question why. Why must we try to devise strategies that attempt to exploit the different possible output voltages of a converter? As stated in the second chapter, the fundamental challenge in power electronics lies in generating dc and ac voltages using combinations of static devices which can only produce distinct voltage levels. Subsequently, we can design filters to process these voltage waveforms with distinct voltage levels, and produce either steady dc voltage or smooth sinusoidal ac voltages. If a converter comprised of multiple controllable devices can generate multiple distinct voltage levels, this can be used to produce a voltage pattern that is much closer to the final desired voltage (ac or dc). By bringing the converter output voltage closer to the final voltage, we are decreasing the burden on the filter, as the unwanted components that need to be removed by the filter will decrease.

   To elaborate on this argument, we can compare the operation of the half-bridge dc-ac converter in the previous chapter with the full-bridge dc-ac converter in this chapter. The half-bridge converter can only produce either $+\frac{V_{dc}}{2}$ or $-\frac{V_{dc}}{2}$, and therefore, during both the positive and negative half-cycles, the converter was producing these two voltages. In a positive half-cycle, there should be no need to produce a negative voltage, but in the absence of any other voltage level besides $+\frac{V_{dc}}{2}$ and $-\frac{V_{dc}}{2}$, one finds that it is essential to use this negative voltage. We began this chapter with bipolar PWM where the full-bridge converter produces an output voltage with only two distinct voltage levels $+V_{dc}$ and $-V_{dc}$. The simulation results showed that we were still able to produce an ac voltage. However, it was clear that we were not fully utilizing the capacity of the converter as two other conduction states were not being used at all.

There are two approaches towards developing a switching strategy that uses all the different voltage levels of a converter. The first is to express the voltage output of a converter as a vector which can attain different values. In the case of a full-bridge converter, with only three possible output levels and one of them being zero, the vector representation was quite simple with the vector being able to attain two values on a single line with zero at the center. In general, each value of the vector can correspond to one or more switching combinations. For the full-bridge converter, the $+V_{dc}$ and $-V_{dc}$ vectors correspond to unique switching combinations but the zero vector can be achieved using two possible switching combinations. With such a vector representation, one can form a mapping between any switched voltage sequence and the vectors that are needed to achieve such an output voltage.

Using vectors sorted in a vector table is used very often for complicated topologies where the output voltage can have several distinct voltage levels. This will be described in the next chapter. However, a simple strategy for PWM so far has been comparing a duty ratio or a modulation signal with either a sawtooth or triangular carrier waveform. Due to the widespread use of this PWM strategy, almost every commercially available microcontroller specially designed for power electronics applications has multiple PWM modules which are extremely easy to configure and use. Therefore, even if we choose a vectorized approach to converter switching, it would be beneficial to attempt to achieve this switching pattern by comparing a modulation signal with a carrier waveform, as from a hardware implementation perspective, the final solution on a microcontroller might be simpler with in-built PWM peripherals than directly using vectors.

Over the past few decades, extensive research has been published in the domain of PWM strategies for converters. From simple converters such as full-bridge converters to advanced multi-level converters, researchers have reported a number of novel switching strategies that provide several benefits such as decreased switching losses, decreased harmonics, decreased dc bus voltage requirements and many more. It is impossible to cover the research reported so far. However, what has been done in this chapter, is to provide the reader with an algorithm which can be used in a systematic manner to devise a switching strategy for a desired switched voltage pattern. As an example, unipolar PWM and phase-shift PWM are chosen which result in switched voltage patterns as shown in Fig. 4.9. These voltage patterns are extremely popular in power electronics applications, and for that reason have been chosen for detailed analysis and simulation.

Unipolar PWM essentially is a strategy that uses voltage levels of a particular polarity to achieve an output voltage of the same polarity. In dc-dc converters, if we wish to achieve an output voltage with a positive average, one needs the converter to produce output voltages that have positive values. In the case of a full-bridge converter, our only options are the $+V_{dc}$ vector and the zero vector. Conversely, if we wish to produce an output voltage with a negative average, a full-bridge converter offers only the $-V_{dc}$ vector and the zero vector. Therefore, to be able to generate switching pulses for a full-bridge converter such that it is flexible in operation and can generate outputs that have either positive or negative averages, we must arrive at a strategy that achieves all conditions. Such a strategy can be arrived at in a step-by-step manner as demonstrated, by considering each operation mode successively, and choosing carrier waveforms such that the desired switching pulses are obtained.

Once this iterative approach to designing a PWM strategy was demonstrated for a dc-dc converter, the strategy could also be applied to a dc-ac converter to achieve unipolar PWM. The prime difference between the dc-dc converter and the dc-ac converter is the nature of the carrier waveform, as for dc-dc converters a sawtooth carrier waveform is used, while for dc-ac converters a triangular carrier waveform is used. PWM strategies are usually found confusing by many newcomers to power electronics, as it may seem like no clear mathematical formula exists. However, this is merely a part of the challenge and the beauty of power electronics, as power converters can be synthesized in a number of different ways. While designing a power converter, one merely needs to understand how energy is extracted from an input and transformed to the final output, while keeping in mind that since the converters use only static non-linear devices, one will need to use passive filters in the form of inductors and capacitors to store energy intermittently. Therefore, depending on how many static controllable devices are being used, one can determine the number of conduction modes possible, and the switching strategy that needs to be devised must merely fully utilize the conduction modes possible in the most optimal manner.

The primary purpose behind this section was merely to reiterate to the reader that designing and controlling power converters is not completely random or heuristic, but can be a systematic process. The reader is encouraged to think of ways in which existing power converters can be modified, and how that would change the conduction modes possible, and also which of the conduction modes would be classified as forbidden. What results is similar to a puzzle, where one can devise switching patterns to achieve these conduction modes, so as to result in a desired output voltage. Besides the switching of the converter, the power electronics engineer must also design passive filters to remove unwanted components from the final output. The presence of passive filters comprising inductors and capacitors has an implication on the current flowing through different parts of the converter, due to which it will become necessary to insert power devices such as diodes to allow the current to freewheel.

4.7 Conclusions

This chapter dealt with the converter that is considered to be the simplest yet fully functional module for power electronics applications. While the half-bridge dc-ac converter is an adaptation that allows for simple ac grid integration for low power applications, the full-bridge converter can be used for medium to high power applications. For this reason, besides simulations, this chapter also focussed on an analytical vector representation of the converter output voltage. It is extremely important that the reader has fully digested the concept of how the converter voltage is represented as a vector that can attain distinct values equal to the distinct output voltages that can be produced by the converter. The next chapter will utilize this concept for the three-phase converter.

As with other chapters, this chapter begins with examining all the conduction modes possible with the full-bridge converter. As the full-bridge converter comprises of two converter legs sharing the same dc bus, it is still forbidden to turn ON both controllable devices in a single leg simultaneously, as this would short-circuit the dc bus. However, the greater number of devices results in a greater number of conduction modes possible. Though there exists a constraint between how the two controllable devices in a single leg can be switched, there exists no constraint between the two legs of the full-bridge converter. As a result, in addition to the conduction modes that produce $+V_{dc}$ and $-V_{dc}$ as the converter output voltage, there are two additional conduction modes that produce a zero output voltage. Therefore, the full-bridge converter is capable of producing three distinct levels in the output voltage.

The concept of vector representation of the converter output voltage was introduced with the full-bridge converter due to the fairly simple representation possible with only three distinct voltage levels. With each vector representing one or more switching combinations, we examined how it is possible to map a desired switched voltage pattern to a vector transition diagram. Such a vector transition diagram can be utilized for generating switching pulses for a converter in a hardware implementation using a microcontroller, where each vector can be stored in a table along with the corresponding switching combination. Though it can be implemented for the full-bridge converter, an implementation using voltage vectors is deferred until the next chapter, when the vector representation will be fairly more complex and therefore, will serve a better example. However, the reader should appreciate how convenient vector transition diagrams are for generating switching pulses for any desired converter output voltage.

Though vector based modulation strategies are well established and especially so for complex converter topologies, in many cases, a PWM strategy based on a comparison between a duty ratio or a modulation signal and a carrier waveform is preferred due to the ease of implementation with a microcontroller. This is also the case of some multi-level converters as will be described in a later chapter, where significant modifications need to be made to the carrier waveforms to continue with the normal comparison strategies. This is due to the fact that though a vector based implementation is quite simple and can be visualized graphically for greater clarity, using a vector based implementation needs the specialized use of timed interrupts that can be a bit complicated in a hardware implementation. In comparison, with the traditional carrier-modulation comparison technique, the only modification necessary is with the carrier waveforms, and most microcontrollers allow for significant flexibility with configuration of PWM peripheral modules.

To be able to generate a strategy to compare carrier waveforms with a modulation signal to generate switching pulses can be fairly confusing to a newcomer to power electronics. In this chapter, several examples have been presented to demonstrate how for a given desired switching voltage pattern, the carrier waveforms can be chosen and the comparison logic can be formulated. Using analysis and simulation, unipolar PWM for dc-dc and dc-ac systems have been demonstrated which describe how conveniently such a strategy can be formulated and implemented. To add a twist to the narrative, the phase-shift PWM strategy has also been described. Though it is used in dc-dc applications, it is primarily used in applications that use high-frequency transformers, and therefore, the applied voltage needs to be of a pseudo-ac nature. Using analysis and simulations, the modifications in the carrier waveforms as well as the comparison algorithm have been demonstrated to result in a phase-shift PWM.

The important contribution of this chapter is to provide the reader an ability to examine a power converter topology in an analytical manner. We began the chapter with examining the conduction modes, which is essential to determine the number of distinct output voltages of the converter. With this knowledge of the distinct output voltages of the converter, we can synthesize switched voltage patterns that utilize the different possible converter output voltages in the most optimal manner while achieving the desired final output voltage. To elaborate on this approach, we chose a few popular examples. We examined how a dc-dc converter can be switched for bidirectional four-quadrant operation, and how it can be switched using unipolar PWM to produce an output voltage with either a positive average value or a negative average value. We also examined how a dc-ac converter can be switched using unipolar PWM to produce an output voltage with a sinusoidal envelope. In all these examples with unipolar PWM, it was very clear that using only select voltage levels resulted in a similar output voltage with lesser unwanted harmonics.

The highlight of this chapter has been the flexibility of PWM strategies as well as the strong analytical approach available to synthesizing PWM strategies. Most newcomers to power electronics are frustrated by the lack of mathematical formulas in developing PWM strategies. However, one must always keep in mind that any power electronic system is a non-linear system for which it is difficult to formulate a precise mathematical approach. Instead, what can be proposed is a strong analytical technique based on understanding the capabilities of a converter, understanding the final requirements of the converter, and thereby devising a strategy to fulfil the final requirements as closely as possible by using the converter's capability in an optimal manner. This strategy will be further extended in the next two chapters.

References

1. *VS-GT75YF120NT 4 Pack IGBT Module* (Vishay Semiconductors, 2021)
2. *F4-75R06W1E3 4 Pack IGBT Module* (Infineon Technologies, 2013)
3. *F4-100R17N3E4 4 Pack IGBT Module* (Infineon Technologies, 2019)

Three-Phase Converters

5

5.1   Introduction

In the previous chapter that dealt with the full-bridge converter, we introduced the concept of representing the converter output voltage as a vector, and subsequently created a vector diagram based on all the possible values that the vector can attain. This established a systematic procedure for analyzing a power converter—examine all possible conduction modes of the converter, determine the output voltage for each conduction mode and create a vector diagram using these voltage values. Though the full-bridge converter is fairly popular for several applications ranging from dc-dc power supplies to single-phase dc-ac converters, a vast number of industrial applications are multi-phase and require high power rating converters for which these full-bridge converters may not be suitable. In this chapter, we examine a basic three-phase converter that has the potential for high power applications.

The chapter begins with an overview of three-phase systems. For readers who may not have a background of power systems and machines, the chapter describes in an abstract manner how three-phase voltages are produced by three-phase generators. The chapter also describes some of the popular connections of three-phase systems by describing the star and delta connections. A detailed treatment of three-phase systems is not covered in this chapter, as the primary objective of this book is to describe switching strategies for power converters, rather than a full system analysis. The objective behind the introduction to three-phase systems is to describe how depending on the star or delta connection of the three-phase system, line-line voltages or line-neutral voltages can be specified. These definitions of voltages will become the specifications of the voltages that the converter is required to produce.

Power electronic converters for three-phase and multi-phase systems in general can be realized in several different ways along with many innovative designs for special applications. However, the chapter presents one of the simplest and also a fairly popular topology of a three-phase converter. Similar to the full-bridge converter in the previous chapter, the

© The Author(s), under exclusive license to Springer Nature Switzerland AG 2024   129
S. V. Iyer and M. N. Aalam, *Switching Strategies for Power Electronic Converters*,
Synthesis Lectures on Power Electronics, https://doi.org/10.1007/978-3-031-41405-3_5

three-phase converter uses the single converter leg consisting of two controllable devices and their associated anti-parallel diodes as the fundamental building block. Using this fundamental building block, one can mathematically compute the total possible conduction modes for the converter. Using an intuitive method, the chapter will list all the conduction modes for the three-leg converter. The reader should take note of the approach used in listing the conduction modes, as this systematic approach can be used for more complex topologies that would otherwise be a bit confusing.

After presenting the topology and conduction modes of the three-phase converter, a basic simulation result using sine-triangle PWM will be presented. This simulation result will present the output currents and output voltages produced by the converter. Though the sine-triangle PWM is a very popular approach towards generating switching signals for a three-phase converter, it does not throw light on how the different conduction states of the converter are used. In the previous chapter on the full-bridge converter, due to the simplicity of the converter, it was possible to implement modified sine-triangle strategies for different converter vector transitions. However, in the case of a three-phase converter, such modifications are not very easy to conceive. Furthermore, to be able to use vector transitions for a three-phase converter, we need to define a single vector that represents the output of a three-phase converter.

To represent the output of a three-phase converter as a single vector, the chapter introduces a very popular mathematical transformation used originally in the domain of machine analysis, namely, the Clarke's transformation. The chapter describes how a three-phase voltage can be represented as a set of rotating vectors. Clarke's transformation is merely a transformation that produces another three quantities through equations involving the three-phase voltages along with static constants. The Clarke's transformation is extremely simple to implement in real-time in any microcontroller or equivalent hardware controller platform. Using rotating vectors, it is shown that the Clarke's transformation results in a pair of rotating vectors that form a quadrature pair. As a result of the transformed quantities resembling a quadrature pair of rotating vectors, their instantaneous values can be plotted in a complex plane. The complex quantity that results, resembles a rotating vector in this new reference plane, and can be used as the vector representation of the three-phase quantity.

Following this description of the vector representation of a complex quantity, the chapter describes how the different outputs of the converter corresponding to every possible switching combination can be transformed using Clarke's transformation. The chapter will present tables of computations of the voltage vectors for every possible switching combination, which the readers are encouraged to verify on their own. As a result of this computation, the outputs of the three-phase converter can be represented as a vector diagram. With this vector diagram, we now have a visual representation of the outputs that a three-phase converter is capable of generating. The vector diagrams have been generated using Python code which the reader is encouraged to review and modify for their own applications.

The chapter describes how the vector diagrams of the three-phase converter can be used to generate switching signals using Space Vector Pulse Width Modulation (SVPWM). The chapter describes how the three-phase output voltage that is desired at any given instant can be transformed using the Clarke's transform, and the resultant complex vector can be superimposed on the vector diagrams. The chapter describes in detail the algorithm which can be used to choose the converter voltage vectors that can be used to realize this required output voltage. Moreover, the chapter also describes the details of the implementation by presenting a method to calculate the time periods of the vectors, besides also deciding the sequence of vectors to be applied over a switching cycle.

The chapter concludes with a simulation of the three-phase converter controlled by SVPWM. Besides the simulation results, the chapter also describes how the vector diagrams of the converter can be used to determine the capability of the converter. This approach is a very convenient visual tool to understand if the converter is suitable for a particular application, or whether the required outputs are beyond the capacity of the converter. The chapter covers a considerable amount of theory and uses Python code to support the theory presented. The reader is encouraged to solve equations that are presented in this chapter along with executing and modifying the simulations that are used to make the theory clearer.

5.2   Three-Phase Systems

Three-phase systems came into being in the 1880s due to the independent efforts of several engineers and researchers with the most notable among them being Nikola Tesla. Over those early years, the pioneering engineers put forth three-phase motors, generators, transformers and also distribution and transmission systems. Using these demonstrations, it was clear that three-phase systems allowed for greater power handling than equivalent conventional single-phase two-wire systems. In this section, we will examine a few basics of three-phase systems which can also be extended in general to polyphase systems with more than three phases for specialized applications.

In order to ease our understanding of three-phase ac systems, let us briefly revisit single-phase ac systems. A single-phase ac voltage is produced by a single-phase ac generator and is made available across two terminals with the following well-known expression:

$$v = V_{pk} \sin(\omega t + \phi) \tag{5.1}$$

In the above expression, V_{pk} is the peak of the ac voltage and $\omega = 2\pi f$ is the angular frequency in rad/s while f is the frequency in Hertz. Though the trigonometric expression is a sine function in the above expression, it could also be a cosine function. ϕ is the phase angle of the ac voltage which accounts for the fact that the ac voltage may not be zero at time $t = 0$.

Without going into the details of machine and generator design, let us describe briefly how such a voltage is produced by the generator. Though the operation of machines and generators vastly differs, at a fundamental level, they are merely electromagnetic machines that work on the principle that the coupling of two magnetic fields produces a mechanical force (motors) or that a conductor that moves in a magnetic field experiences a voltage induced across it (generators). To be able to make the explanation a little clearer, let us choose a permanent magnet synchronous motor as an example as shown in Fig. 5.1. In the case of a permanent magnet synchronous generator, the rotational part of the generator called as the rotor, is a permanent magnet with two or more poles (P) $(P = 4$ in Fig. 5.1) that produce a magnetic field. The stationary part of the generator called as the stator has conductors wound in slots (not shown in Fig. 5.1) within the stator to form coils with numerous turns. The exact nature of the windings in the stator can be quite complicated, as these windings are distributed over the circumference of the stator. However, for a simple explanation such as the one we are undertaking, we need only take into account that the winding in the stator can be thought of as a coil with a certain number of turns and therefore, one can imagine a coil with a number of turns on a single plane passing through the center of the generator. This coil naturally will terminate at two terminals which are called the stator winding terminals and with which the generator can be connected to another load or to any other system.

When the rotor is at rest, the magnetic field produced by it is steady and constant and though it associates with the stator, it has no inductive effect on the stator windings, as for any electromagnetic induction to occur across a conductor, the magnetic field associated with the conductor must change. When the rotor is coupled to a prime mover such as gas turbine, diesel engine or even a wind turbine, it begins to rotate. With this rotation, the magnetic field associated with the stator winding coil starts changing. According to Faraday's law, the emf induced in the stator winding coil will be directly proportional to the rate of change of magnetic flux linkage of the winding. The magnetic flux linkage of the stator winding is merely the product of the flux associated with (passing through) the winding and the number of turns of the winding. Therefore, the faster the rotor rotates and the greater the number of turns in the stator winding, the greater will be the emf produced in the stator winding.

The rotor has a certain maximum speed of operation which is determined by the mechanical structure of the rotor. With respect to this maximum speed, one will usually be given a rated speed by the manufacturer of the generator which is a speed that the rotor can safely rotate at on a continuous basis and which will utilize the mechanical capacity rotor to the

Fig. 5.1 Conceptual depiction of a permanent magnet ac generator

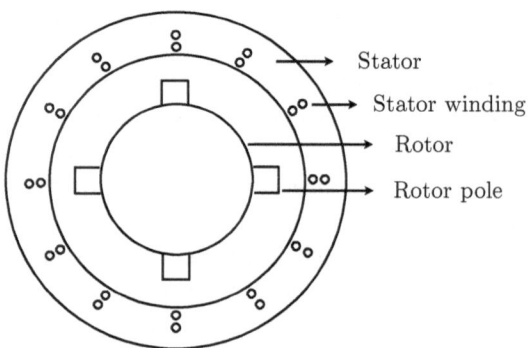

fullest. If the prime mover can be regulated to rotate the rotor at this rated speed, this will result in the voltage induced in the stator windings to be of a frequency equal to the rated frequency. Naturally, this rated frequency is usually the grid frequency which is either 50 Hz or 60 Hz depending on the geographic location. The rotor poles as well as the stator winding are so designed that at this rated speed of rotation of the rotor, the ac voltage induced in the stator winding will have a Root Mean Squared (RMS) value equal to the rated voltage of the generator [1]. This rated voltage of the generator is merely a voltage at which we wish to interface the generator to an ac system and therefore can be one of any of the typical values of ac voltage found in ac systems such as 120, 240, 400, 690 V and many more.

With this abstract description, we have managed to understand how a single-phase ac voltage can be produced by an ac generator. The question then arises, how can such a generator be converted into a three-phase ac generator. Before we examine the specific case of a three-phase generator, let us look at the impact of having multiple stator windings. In the previous case, we examined the case of a single coil wound on a stator. We could have multiple coils wound on the stator. This would imply a larger stator as a greater circumference would be needed to accommodate the increased number of slots and conductors. However, as the rotor rotates, it will induce an emf in all the stator windings. What would determine the difference between the emfs induced in the different windings? Answer: their physical position on the stator. As before, if we ignore the fact that a stator winding is distributed around the circumference of the stator and only consider a winding as a symbolic coil on a plane that passes through the center of the generator, it is clear that we have plenty of freedom with respect to positioning the windings.

As an example, we could superimpose the two winding coils such that they are essentially parallel-connected windings. In that case, the voltage generated across the terminals of the two windings will be identical. Alternatively, the plane of the second stator winding could be perpendicular to the existing stator winding. Now the two stator windings are physically separated and will experience a different magnetic flux linkage and rate of change of flux linkage at the same instant of time. Therefore, the voltages generated at the terminals of the stator windings will no longer be identical, but will be phase-shifted with respect to each other. The phase-angle between the voltages will be the electrical angle of separation between the two stator winding coils. To understand the concept of electrical angle of separation, one must take into account the number of poles (P) of the rotor.

The permanent magnet rotor can have any number of magnetic poles as long as this number is even, as magnetic poles are always present in North-South pole pairs. In case the number of magnetic poles is greater than 2, this results in a difference between the physical span of the machine versus the electrical span over a cycle of the voltage produced. For example, if there were 6 magnetic poles (3 pairs of North-South poles), a single complete rotation of the rotor will produce three cycles of electrical voltage. This is due to the fact that when a coil passes from one South pole to one North pole and further to the next South pole, this will result in a complete cycle of change of the magnetic flux linkage of the stator coil, and therefore will induce one cycle of voltage across the stator coil terminals. However, during one complete rotation of the rotor, such a transition will occur thrice resulting in three full ac voltage waveforms. In short:

$$\text{Electrical angle} = \text{Mechanical angle} \times \frac{\text{No of Poles}(P)}{2} \qquad (5.2)$$

With this background, if the stator winding coils were in planes that were perpendicular to each other, this would imply that the phase angle between the voltages generated at the two windings will be $90° \times \frac{P}{2}$. For our specific case of a 4 pole generator, this implies that the voltage waveforms are in reality 180° phase shifted with respect to each other. If we take into account the fact that a single cycle of ac voltage spans over 360° if one considers both positive and negative half-cycles, we could think of multiple voltages that are equally phase shifted across one cycle. An example was the two stator windings above, where the phase shift between them is equal to $\frac{360°}{2}$. We can consider a stator with three windings such that the phase angle between the winding voltage waveforms would be $\frac{360°}{3} = 120°$. To achieve such voltages, the windings would need to be physically displaced in the stator circumference by $120° \times \frac{2}{P}$.

The advantage of such a multi-phase machine is greater power at a comparatively lower size. The presence of three stator windings will naturally require greater number of slots to house the winding conductors and therefore will need a larger stator circumference. Due to the presence of three stator windings, the total electrical power produced by the stator will be greater, and therefore the mechanical power in the rotor will also have to be greater, which in turn implies a larger rotor. However, these increases in the size of the machine parts will not be as large as would be the case if we were to design a single-phase ac generator with three times greater capacity. This is one of the greatest advantages of multi-phase systems over single-phase systems, due to which large capacity machines and transmission systems are usually three-phase or in some special applications multi-phase.

With the above description of how three-phase ac voltages can be generated, let us extend the expression for the ac voltage to express three-phase ac voltages:

$$v_a = V_{pk} \cos(\omega t + \phi)$$
$$v_b = V_{pk} \cos(\omega t - 120° + \phi) \qquad (5.3)$$
$$v_c = V_{pk} \cos(\omega t - 240° + \phi)$$

The three-phases are called phase a, phase b and phase c, though many other terminologies also exist. One such popular terminology arises from the color of wires used for the three different phases resulting in phase Red (R), phase Yellow (Y) and phase Blue (B). The one question that arises when we speak of voltages is how the terminals will be connected. In the case of a single-phase ac generator, one of the terminals is designated a live or phase terminal and the other terminal is called the neutral. In the case of three-phase ac generators, with three stator windings, we have two terminals per winding, and therefore, six terminals altogether. How these terminals can be made available results in a few three-phase connections.

Figure 5.2 shows three of the most popular connections of three-phase ac voltages and also three-phase systems in general. The left-most connection in Fig. 5.2 is called a star connection with isolated neutral. One of the terminals of all the stator windings are connected together to form a neutral. However, this neutral connection stays internal to the machine and is not available externally. Therefore, only three terminals a, b and c are available externally. The middle connection in Fig. 5.2 is called a delta connection. In this case, the windings are connected in a chain, and as a result only three terminals a, b and c are available externally without any neutral connection. The right-most winding in Fig. 5.2 is called a star connection with neutral. It is merely an extension of the left-most connection, except that the neutral connection is made available externally in addition to the three phase terminals a, b and c. There are other ways in which one can connect three phase windings, however, for the content that will follow in this chapter, the reader need only be familiar with these connections.

Figure 5.3 shows how the three phase voltages can be expressed as rotating vectors. All three vectors are rotating at an angular speed $\omega = 2\pi f$ and in the counter clockwise direction as indicated. Phase b lags behind phase a by $120°$ while phase c lags behind phase b by $120°$. All three phase voltages have an initial phase angle of ϕ, which is most prominent in the phase a voltage. The projection of each vector on the horizontal axis is the instantaneous

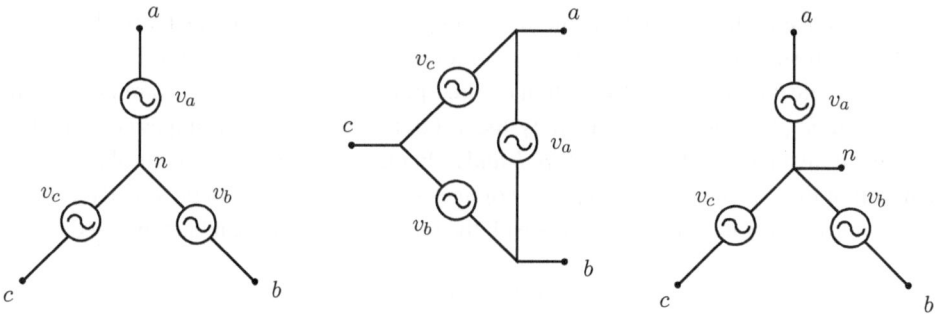

Fig. 5.2 Star and delta connections of three-phase voltages

Fig. 5.3 Three-phase voltages
expressed as rotating vectors

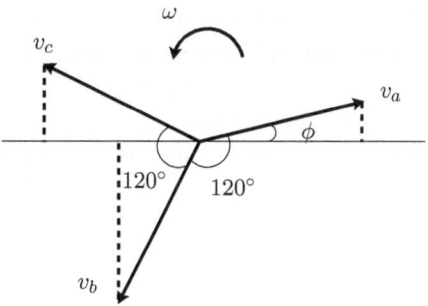

value of the phase voltage due to the phase voltages being cosines of the phase angle. Such vector representations will be used throughout this chapter and also to a certain extent in the next chapter. In vector representations, the speed of rotation ω as well as the direction will usually not be specified as it is assumed that all vectors are rotating at the same speed equal to the angular frequency of the ac system, and the direction of rotation can be inferred from the sequence of the vectors. The reader is encouraged to verify that the representation of vectors in Fig. 5.3 corresponds to their equations above [2].

Additionally, it is also important to speak of balanced versus unbalanced three-phase systems. When three-phase systems are such that the three voltages produced have the same RMS value, and when they are phase shifted with respect to each other by 120°, with phase b lagging behind phase a and phase c lagging behind phase b, such a system is called a balanced three-phase ac system. Both the equations above as well as the vector representation in Fig. 5.3 belong to a balanced three-phase system. In contrast, if the RMS magnitudes of the three voltages are not equal, or if the phase angle between them is not 120°, such a system is called an unbalanced three-phase system. A vast number of domestic and consumer appliances in households are single-phase appliances while three-phase appliances and machines are more commonly found in industries and factories. Therefore, when a three-phase ac system is supplied to a household, the appliances in the household are roughly divided across the three phases. However, this inevitably results in a certain degree of unbalance, as it would be impossible to equally distribute appliances across three phases. Such unbalance eventually creeps into the larger system resulting in unbalances both in the low voltage distribution system and also at times in the transmission system.

To conclude on this basic introduction to three-phase systems, we will examine one other aspect of three-phase ac voltages, namely, the difference between the voltages generated and the voltages available at the terminals a, b and c. In the star with isolated neutral connection in the left of Fig. 5.2, the generated phase voltages v_a, v_b and v_c will not be directly available at the external terminals a, b and c. Instead, the terminal voltages can be expressed as:

$$v_{ab} = v_a - v_b$$
$$v_{bc} = v_b - v_c \qquad (5.4)$$
$$v_{ca} = v_c - v_a$$

The terminal voltages above are also called line-line voltages, as each external terminal is a line in a three-phase system. The reader is encouraged to use the trigonometric expressions for v_a, v_b and v_c, to compute the expressions for the line-line voltages v_{ab}, v_{bc} and v_{ca}. The reader will find that the line-line voltages have a RMS value which is $\sqrt{3}$ times the RMS value of the phase voltages. For the delta connection in the middle of Fig. 5.2, the generated phase voltages can be seen to be equal to the line-line voltages as follows:

$$v_{ab} = v_a$$
$$v_{bc} = v_b \qquad\qquad (5.5)$$
$$v_{ca} = v_c$$

Therefore, for delta connected systems, the line-line voltages have the same RMS magnitude as the phase voltages. For the last case of the star connection with neutral, one might be tempted to equate line-line voltages with phase voltages as the neutral is available as an external terminal. However, the neutral connection is not usually used for measuring line-line voltages, and therefore the line-line voltages of a star connection with neutral are the same as the line-line voltages for a star connection with isolated neutral.

This section provided a quick introduction to three-phase systems. Though three-phase systems can be studied in far greater detail, for the later sections, the above discussion should provide the reader with a quick overview of the context. The section should be useful to a reader who has a purely power electronics background and has forgotten basic power systems, as the section describes in a brief and abstract manner how a three-phase system comes into being. Furthermore, once such a three-phase system comes into being, it is essential that we learn of the popular three phase connections, namely star and delta. Three-phase quantities (voltages and currents) can be expressed as vectors (sometimes also called phasors), and this visual depiction provides all the information about the three-phase system as contained in the mathematical expressions. With this background, we will examine in the next section, a power electronics converter that can be interfaced to such a three-phase system.

5.3   Three-Phase Converter Topology and Conduction Modes

In the previous section, we examined the basics of three-phase systems, a few of the popular connection types and a simple vector representation. In this section, we will introduce the topology of a basic power converter that can be interfaced to three-phase ac systems, and subsequently, examine the various conduction possibilities. This approach is similar to that in previous chapters, and a quick and visual approach to understanding the operation of a power converter. It should be noted that three-phase converters can be conceived in a number of different ways, and the topology being examined in this section could be said to be one of the simplest and most basic.

Figure 5.4 shows the topology of a three-phase converter connected to the three terminals of three resistor-inductor loads connected in star with isolated neutral. The three-phase converter of Fig. 5.4 is also called a three-leg converter due to the fact that it is comprised of three legs each comprised of two controllable power devices with their associated anti-parallel diodes. The three-phase converter shown can be used in a number of applications, in motor drives, for producing three-phase ac power supplies, for interfacing renewable energy supplies and many more. Since we are only concerned with the switching of the converter, let

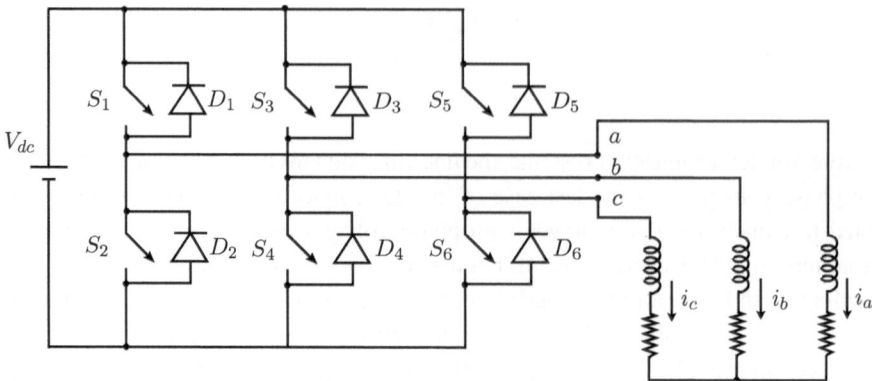

Fig. 5.4 Three-leg converter interfaced to three phase $R - L$ loads

us consider the case of the converter supplying a three-phase R-L load as shown in Fig. 5.4. After the examination of the full-bridge converter in the previous chapter, it is clearly evident how a single leg can be used as a building block for synthesizing complex converters. If one imagines multi-phase systems with N number of phases, one could think of the simplest power converter for such a N-phase system to have N converter legs.

As with our analysis in the previous chapters, we can state that since the three-leg converter consists of three converter legs, the forbidden conduction modes are those in which both controllable devices of one or more legs are conducting simultaneously as these will short-circuit the dc bus. Similar to the full-bridge converter, it is clear that the legs can operate independently with respect to each other, and these will not produce forbidden states. For the sake of simplicity, the operating modes will only depict the controllable devices, as it should be assumed by the reader that if the current were such that a controllable device cannot conduct, the anti-parallel diode across the controllable device will conduct instead. Due to the greater number of conduction modes, the three-leg converter is a good case for determining the number of conduction modes possible.

For a single converter leg, it is quite clear that there are two conduction possibilities, either the upper device or the lower device. As there is no restriction between the operation of the legs with respect to each other, the total number of combinations possible for the three-leg converter will be:

$$2 \times 2 \times 2 = 2^3 = 8 \tag{5.6}$$

In general, for a N-leg converter with each leg comprising of two controllable devices with their associated anti-parallel diodes, the total number of combinations will be 2^N.

These combinations are depicted in Fig. 5.5. For simplicity, the entire circuit of Fig. 5.4 has not be shown. However, the reader is encouraged to draw some of the conduction modes with the rest of the circuit of Fig. 5.4 and draw the conduction loops. From Fig. 5.5, the reader can quickly note that two of the operating modes will result in the converter producing a zero

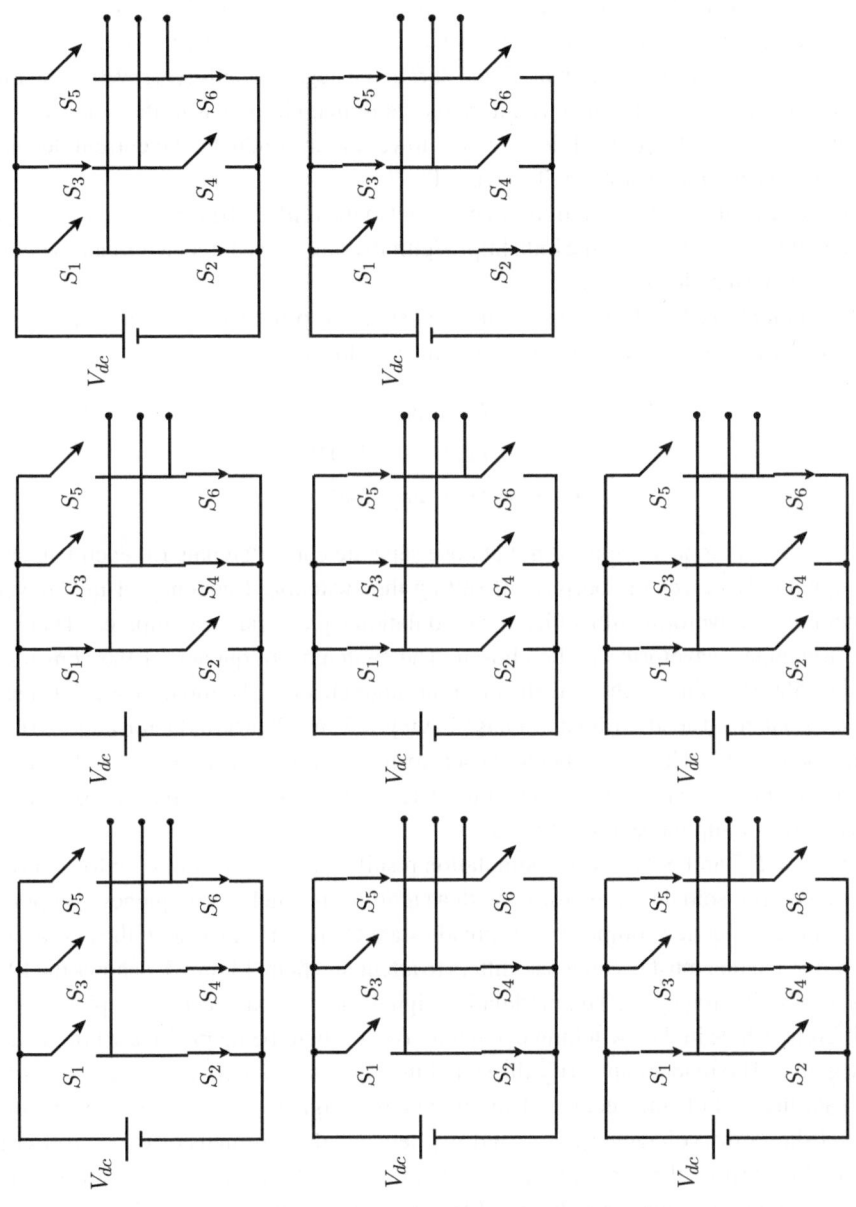

Fig. 5.5 Three-leg converter conduction modes

output voltage—when all the upper devices are conducting or when all the lower devices are conducting. In the other operating modes, due to the fact that the system has no neutral connection, all other conduction modes involve loops formed between phases. However, in all cases, one can formulate a general rule. When the upper controllable device conducts, the current increases in magnitude during the positive half-cycle, while when the upper diode conducts, the current decreases in magnitude during the negative half-cycle. When the lower controllable device conducts, the current increases in magnitude but in the negative sense during the negative half-cycle, while when the lower diode conducts, the current decreases in magnitude during the positive half-cycle [3].

The basic bipolar PWM simulation can be found in the folder `three_leg_sine_pwm` within the folder `chapter5` at the link: https://github.com/opensourceelectrical/switching-strategies-for-power-electronics.

In this simulation, the three legs of the converter are provided with three modulation signals that form a three-phase balanced system as follows:

$$m_a = 0.7\cos(\omega t)$$
$$m_b = 0.7\cos(\omega t - 120°) \tag{5.7}$$
$$m_c = 0.7\cos(\omega t - 240°)$$

In terms of carrier signals, we need not choose separate carrier signals for each leg, as the carrier signal only serves the purpose of setting the switching frequency of the converter and generating a waveform with which the modulation signals can be compared. Therefore, a single triangular waveform can be chosen. The switching frequency of the converter is chosen as 2500 Hz to make the waveforms a bit more clearer. The three-phase R-L load is chosen as a balanced load with each phase having a 1 millihenry inductor and a 10 ohm resistor. The dc bus voltage has been chosen to be 24 V for the purpose of illustration, though, for practical systems, one would either need a much higher dc bus or would need a transformer to step-up the ac output voltages.

Figures 5.6, 5.7 and 5.8 show the simulation results. Figure 5.6 shows the load currents to be balanced sinusoidal ac waveforms with a significant switching frequency component. This switching frequency component as already stated before is intrinsic to the operation of the converter and one must use low pass filters to attenuate them. Figure 5.7 shows the PWM process for a switching cycle. The modulation signals being a balanced three-phase system have different values in the switching cycle with two of them being positive while the third being negative. The reader can verify the switching signals by zooming into the waveforms at the exact instants of intersection of the modulation signals and the carrier waveforms. Figure 5.8 shows the voltages v_{an}, v_{bn} and v_{cn} between the terminals of the converter and the isolated neutral of the three-phase load. Quite strangely, these voltages are not at all similar to the bipolar switching voltages of the converters in the previous chapters. Though this might appear at first glance as a bug, this can be explained once one writes the network equation for the system [4].

Fig. 5.6 Three-leg converter output currents

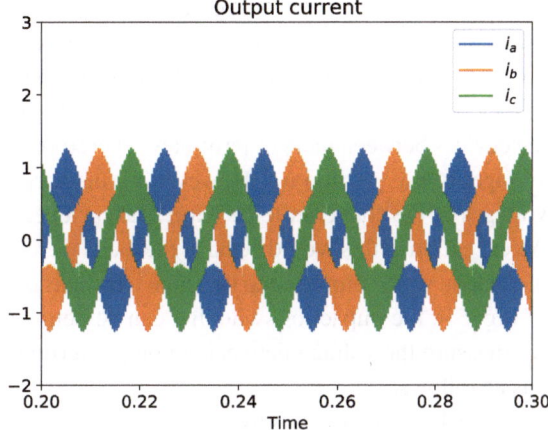

Fig. 5.7 PWM for three-leg converter

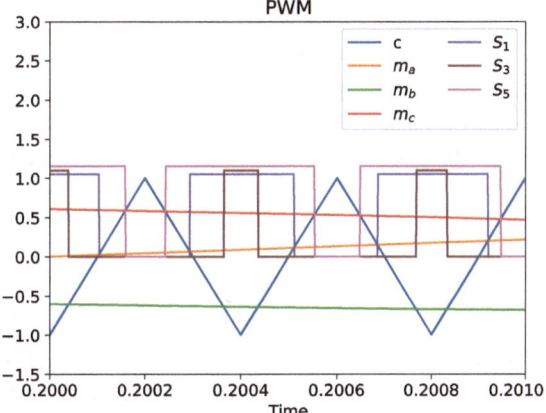

Fig. 5.8 Three-leg converter output voltages

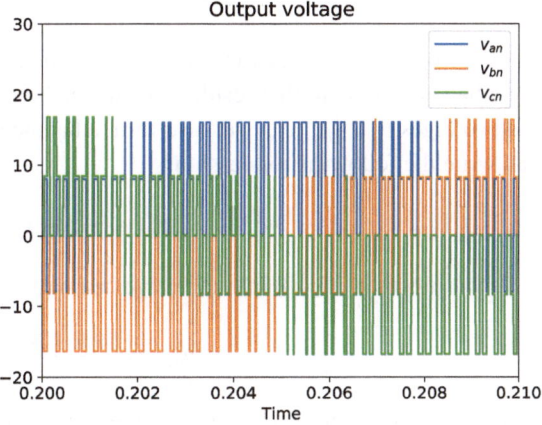

In the case of the full-bridge converter of the previous chapter, the output voltage of the converter appears across the two legs and can be $+V_{dc}$, $-V_{dc}$ or 0. In the case of a three-leg converter, with the three legs of the converter operating fairly independently, the output voltages across the three phases can be determined with respect to the switching of the devices between pairs of terminals. Therefore, one could assign similar $+V_{dc}$, $-V_{dc}$ or 0 voltages to v_{ab}, v_{bc} and v_{ca}. However, for the purpose of analysis, one would like the output voltages of the converter to be expressed on a per-phase basis, namely as v_{an}, v_{bn} and v_{cn}. An exception to this would occur if the loads were connected in delta and the three terminals of the load were connected to the three terminals of the converter. For the three-leg converter of Fig. 5.4, the impact of switching can be defined with respect to each output terminal, if we measure the voltage between an output terminal and the negative terminal N of the dc bus as follows:

$v_{aN} = +V_{dc}$, when S_1 is ON, S_2 is OFF
$v_{aN} = 0$, when S_1 is OFF, S_2 is ON
$v_{bN} = +V_{dc}$, when S_3 is ON, S_4 is OFF
$v_{bN} = 0$, when S_3 is OFF, S_4 is ON
$v_{cN} = +V_{dc}$, when S_5 is ON, S_6 is OFF
$v_{cN} = 0$, when S_5 is OFF, S_6 is ON

For the system of Fig. 5.4, the following three loop equations can be written:

$$v_{aN} - v_{an} - v_{nN} = 0$$
$$v_{bN} - v_{bn} - v_{nN} = 0 \qquad (5.8)$$
$$v_{cN} - v_{cn} - v_{nN} = 0$$

The voltage v_{nN} is an arbitrary voltage between two floating terminals and therefore, it is best that we eliminate it. This can be achieved by taking the sum of the above three equations:

$$(v_{aN} + v_{bN} + v_{cN}) - (v_{an} + v_{bn} + v_{cn}) - 3v_{nN} = 0 \qquad (5.9)$$

In the above equation, due to the three-phase load being balanced, if the modulation signals are a balanced system that result in a balanced three-phase current flowing in the load, the load voltages v_{an}, v_{bn}, v_{cn} will also form a balanced system. Being a balanced system, one can write the following expression:

$$v_{an} + v_{bn} + v_{cn} = 0 \qquad (5.10)$$

If we substitute (5.10) into (5.9), we can write the following expression for v_{nN}:

$$v_{nN} = \frac{v_{aN} + v_{bN} + v_{cN}}{3} \qquad (5.11)$$

With this expression for v_{nN}, we can rewrite the equations in (5.8) as follows:

$$v_{aN} - v_{an} - \frac{v_{aN} + v_{bN} + v_{cN}}{3} = 0$$

$$v_{bN} - v_{bn} - \frac{v_{aN} + v_{bN} + v_{cN}}{3} = 0 \tag{5.12}$$

$$v_{cN} - v_{cn} - \frac{v_{aN} + v_{bN} + v_{cN}}{3} = 0$$

The above equations can be simplified to produce expressions for v_{an}, v_{bn} and v_{cn} as follows:

$$v_{an} = \frac{2}{3}v_{aN} - \frac{1}{3}v_{bN} - \frac{1}{3}v_{cN}$$

$$v_{bn} = -\frac{1}{3}v_{aN} + \frac{2}{3}v_{bN} - \frac{1}{3}v_{cN} \tag{5.13}$$

$$v_{cn} = -\frac{1}{3}v_{aN} - \frac{1}{3}v_{bN} + \frac{2}{3}v_{cN}$$

Since the voltages v_{aN}, v_{bN} and v_{cN} are either $+V_{dc}$ or 0, it is very clear that the voltages v_{an}, v_{bn}, v_{cn} can have only a few possible values. The reader is encouraged to substitute different possible values into the above equation and verify that v_{an}, v_{bn}, v_{cn} can have the following values—$\frac{2}{3}V_{dc}$, $\frac{1}{3}V_{dc}$, 0, $-\frac{1}{3}V_{dc}$, $-\frac{2}{3}V_{dc}$. If one re-examines the simulation result of Fig. 5.8, it is very clear that the voltages have these unique levels. The reader is further encouraged to plot the switching pulses for each leg and verify that the voltages in Fig. 5.8 are the result of the equations in (5.13). The reader is also encouraged to repeat the simulation while connecting the three-phase load in a delta connection and observing the change in the results.

In this section, we examined a basic three-phase converter comprised of three converter legs that has the capability of supplying a three-phase system. Using our previous knowledge of multiple legs connected across the same dc bus, we determined the number of unique conduction modes that were possible, and subsequently depicted them. We presented a simulation study that enables us to examine in greater detail the conduction modes of the three-phase converter using sine-triangle PWM. We further examined the impact of the isolated system neutral, and modified the equations for the converter phase output voltages. Though the simulation enables us to examine the operation of the three-leg converter, it is clear from the simulation result that we do not have the same control over choosing the exact three-phase output voltage of the converter, but rather produce three independent output voltages. Due to the multi-phase nature of the converter, we can no longer use multiple carrier waveforms to be able to choose the exact multi-phase output voltage that the converter should produce. The next sections will gradually examine an alternate modulation strategy that enables us to overcome this limitation.

5.4   Vector Representation of the Three-Phase Quantities

In the previous section, we examined the topology of a basic three-leg converter that can supply a three-phase system. We presented a basic simulation where each leg is operated independently using sine-triangle PWM with a modulation signal from a balanced set of signals and a triangular carrier waveform. To formulate a switching strategy for the three-leg converter that fully utilizes the the capability of the converter, we must be able to generate a vector representation and subsequently, to examine transitions between the vectors. Before examining the converter, we must first find a convenient vector representation for a three-phase ac quantity, and for this purpose, we will introduce an extremely popular transformation. In this section, we will examine Clarke's transformation that is used extensively in the analysis and control of machines and also for power electronics in general.

The Clarke's transformation or the abc-$\alpha\beta 0$ transformation is a transformation that originated from machine analysis but later began to be applied in the analysis of any three-phase ac system [5–7]. A transformation is a mathematical operation performed on a system that results in an equivalent system that has different properties thereby facilitating a certain form of analysis. In Clarke's transformation, one transforms a three-phase system with phases a, b, c to another system with phases α, β and 0. It is important to note that one does not physically build a machine in the α-β-0 domain as it is merely a mathematical representation. Furthermore, it is also important to note that a transformation should not result in a completely unrelated system, as then, an analysis of the transformed system would not be of any use in understanding the original system. When one is assured that the transformed system would behave identically to the original system, then we can be assured that any conclusions we draw from the transformed system would apply to the original system as well [5].

Since the Clarke's transformation was originally intended for machine analysis, let us consider a three-phase set of currents as follows:

$$
\begin{aligned}
i_a &= I_{pk} \cos(\omega t + \phi) \\
i_b &= I_{pk} \cos(\omega t - 120° + \phi) \\
i_c &= I_{pk} \cos(\omega t - 240° + \phi)
\end{aligned}
\tag{5.14}
$$

The transformed currents in the α-β-0 frame of reference are expressed as:

$$
\begin{bmatrix} i_\alpha \\ i_\beta \\ i_0 \end{bmatrix} = \frac{2}{3} \begin{bmatrix} 1 & -\frac{1}{2} & -\frac{1}{2} \\ 0 & \frac{\sqrt{3}}{2} & -\frac{\sqrt{3}}{2} \\ \frac{1}{\sqrt{2}} & \frac{1}{\sqrt{2}} & \frac{1}{\sqrt{2}} \end{bmatrix} \begin{bmatrix} i_a \\ i_b \\ i_c \end{bmatrix} = \mathbf{T} \begin{bmatrix} i_a \\ i_b \\ i_c \end{bmatrix}
\tag{5.15}
$$

where the matrix $\mathbf{T}$ is the transformation matrix.

The reader is encouraged to substitute the values of i_a, i_b, i_c into the transformation equation and verify that the values of i_α, i_β, i_0 are:

$$\begin{bmatrix} i_\alpha \\ i_\beta \\ i_0 \end{bmatrix} = I_{pk} \begin{bmatrix} \cos(\omega t + \phi) \\ \sin(\omega t + \phi) \\ 0 \end{bmatrix} = I_{pk} \begin{bmatrix} \cos(\omega t + \phi) \\ \cos(\omega t - 90° + \phi) \\ 0 \end{bmatrix} \tag{5.16}$$

As can be observed there is a $90°$ phase difference between i_α and i_β, with i_β lagging behind i_α. The value i_0 is 0 due to the balanced nature of the currents i_a, i_b, i_c. The fact that i_0 is zero has a very special significance in the use of this transformation and therefore, needs some greater explanation.

i_0 is dependent on the sum $i_a + i_b + i_c$ of the three phase currents i_a, i_b, i_c. In the case of a balanced three-phase system, this sum will be zero as can be verified by expanding the trigonometric expressions for the currents. However, this sum will also be zero under certain other conditions even when the three-phase system is not balanced. As an example, in Fig. 5.4, the three-phase load is connected in star with an isolated neutral. In such a case, if the three-phase load were supplied by unbalanced three-phase voltages, the currents flowing through the three phases will not form a balanced three-phase set. However, their sum will still be zero. This is due to the fact that the neutral is isolated. For the sum of the currents to be non-zero in an unbalanced system, a neutral wire must be present and connected to the neutral of the three-phase voltages. The current flowing in the neutral wire will be equal to the sum of the phase currents.

From the above discussion, when the above Clarke's transformation is applied to a three-phase balanced system or to a three-phase system that does not have a neutral connection, the 0 component will be zero. As a result, Clarke's transformation has resulted in a three-phase system being transformed into a pseudo two-phase system. The decrease of one variable presents significant advantages in mathematical analysis that is of great benefit in machine analysis and control design. One can find vast literature on how balanced and unbalanced three-phase systems can be dealt with using Clarke's transformation. However, in the context of this book, our objective is merely to understand switching strategies for converters. Therefore, we can safely assume three-phase systems to be balanced. Using Clarke's transformation on three-phase balanced voltages, one can express them in α-β form as follows:

$$\begin{bmatrix} v_\alpha \\ v_\beta \\ v_0 \end{bmatrix} = V_{pk} \begin{bmatrix} \cos(\omega t + \phi) \\ \sin(\omega t + \phi) \\ 0 \end{bmatrix} \tag{5.17}$$

Figure 5.9 shows the Clarke's transformation performed on the three-phase voltage vectors of Fig. 5.3. In Fig. 5.3, the three-phase voltages were expressed as vectors rotating at an angular speed equal to the angular frequency ω of the grid. It has been assumed that the vectors were rotating in the counter clockwise direction, and furthermore, the phase angle of the voltages was equal to the angle between the horizontal axis and the vector. The instantaneous values of the three-phase voltages were equal to their projections on this horizontal x-axis since the voltages were expressed as cosines of the phase angle. If the reader performs simple vector computations using the three vectors shown for the expressions of v_α and v_β,

Fig. 5.9 Clarke's transformation expressed in rotating vectors

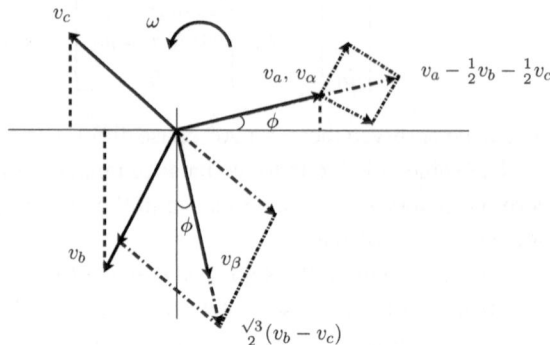

the reader will find the resultant vectors to be as shown in Fig. 5.9. It is quite clear that the vector v_β lags behind the vector v_α by 90°. However, the instantaneous value of the v_α and v_β vectors will be obtained by taking their projections on the horizontal axis as shown in Fig. 5.9 as that is the convention established with defining the instantaneous of any rotating vector.

It is important to note that Fig. 5.9 is a vector diagram where all the vectors are rotating in the counter-clockwise sense at an angular speed ω equal to the angular frequency of the system. Due to the balanced nature of the three-phase voltages, the transformation produces only two vectors v_α and v_β. One can take advantage of this decrease in variables, to define a new voltage variable using the transformed variables:

$$\overline{V} = v_\alpha + j v_\beta \tag{5.18}$$

The above equation combines v_α and v_β into a single complex variable due to the fact that they are in reality vectors that are 90° apart in phase angle, thereby forming an orthogonal pair of vectors. This complex representation of the transformed voltages enables us to represent the transformed voltages in another set of axes called the α-β axes as shown in Fig. 5.10. Following the convention, the α axis is the real axis and the β axis is the imaginary axis. For any given values of v_α and v_β on the α and β axes respectively, we can denote the resultant variable $\overline{V}$ as shown in Fig. 5.10.

Since v_α and v_β vary with time, this will result in the variable $\overline{V}$ resembling another rotating vector. The reader is encouraged to verify that if one maps the vector projections from Fig. 5.9 into the complex plane of Fig. 5.10, the variable $\overline{V}$ also appears to rotate in the counter-clockwise sense at an angular speed ω equal to the angular frequency of the system. As a result of Clarke's transformation, we have represented a set of three-phase voltages by a single rotating pseudo vector $\overline{V}$ which is in reality merely a complex variable synthesized using the result of the transformation. This representation is a great benefit in both analysis as well as control design for three-phase systems. However, it is also important to stress that such a convenient representation is due to the balanced nature of the three-phase quantities.

Fig. 5.10 Complex plane representation of result of Clarke's transformation

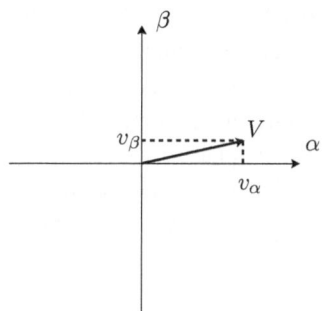

For unbalanced systems, one needs to perform modifications to continue using Clarke's transformation to analyse three-phase systems.

Before we conclude this section, let us examine the physical significance of Clarke's transformation. The theory behind Clarke's transformation is fairly vast and it is impossible to cover it in detail in this book. However, as stated before, Clarke's transformation was originally conceived in the domain of machine analysis, where it can be used to transform a three-phase machine into an α-β-0 equivalent machine. As before, if we assume the machine winding currents i_a, i_b, i_c to be a balanced three-phase set, this will result in i_0 being 0, and the result of Clarke's transformation is merely i_α and i_β. We can now build an equivalent machine with windings through which i_α and i_β flow as shown in Fig. 5.11. If the three-phase machine had three stator windings with N_a, N_b, N_c turns carrying currents i_a, i_b, i_c, the equivalent machine with N_α and N_β turns carrying currents i_α and i_β would be identical in terms of magnetic flux produced and also mechanical capacity, namely torque, speed and power. To find details on winding turns, transformation and machine performance, the reader can refer to [6, 7].

In this section, we examined Clarke's transformation using which a conventional three-phase quantity can be transformed into another three-phase quantity. The transformation has a special implication for balanced three-phase systems, as the result of the transformation is a two-phase system. Using vector diagrams, it was clear that the resultant vectors formed

Fig. 5.11 Physical significance of Clarke's transformation

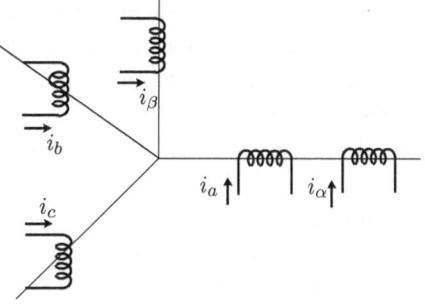

a quadrature pair when the three-phase system was a balanced one. Due to this quadrature nature, the resultant two phases can be represented on a complex plane, and can be combined together as real and imaginary quantities to form a complex variable. This complex variable assumes the property of a rotating vector if one takes into account the changes in the two resultant phases. As a result, a three-phase system has been represented as a single complex variable or a pseudo vector. In the next section, we will examine how this representation can be used to analyse the output of the three-leg converter of Fig. 5.4.

5.5   Vector Representation of the Three-Leg Converter

In the previous section, we examined how Clarke's transformation can be used to transform a balanced set of three-phase quantities into a single rotating vector. In this section, we will use Clarke's transformation to generate vectors for all the possible switching combinations of the three-leg converter of Fig. 5.5 listed in a previous section. With such a representation, we will now have a visual representation of the capacity of the converter and thereafter, be able to synthesize switching strategies such that the converter can produce desired three-phase output voltages. Before reading this section, the reader is encouraged to review the previous section that describes Clarke's transformation. Specifically, the reader should be well-versed with how a balanced three-phase quantity can be transformed into α-β vectors, and how the instantaneous values of these α-β vectors can be mapped in a complex plane to produce a rotating vector.

As already described in a previous section, three-phase systems can be connected in several different ways, with the most popular connections being the star with isolated neutral, the delta, and the star with neutral. For the moment, let us focus on the star with isolated neutral and the delta connections as we are considering balanced systems and therefore, the neutral wire has no significance. With only the three phase terminals a, b, c available to us, we can define the three-phase voltages supplied to these three terminals as either line-line voltages v_{ab}, v_{bc}, v_{ca} or line-neutral voltages v_{an}, v_{bn}, v_{cn}. As an example, if a load or machine were to be connected in delta and a neutral terminal did not exist, the only voltages that one can define are the line-line voltages. On the other hand, for a load or machine connected in star with isolated neutral, it is possible to define line-neutral (phase) voltages as well as the line-line voltages. To be able to fulfil the voltage requirements of a load being supplied by a three-leg converter, we must first define the output voltages of the three-leg converter that will supply this load.

Though some of this discussion has been covered in a previous section, the most relevant material will be repeated for convenience. From Fig. 5.4, the switching output of the converter can be defined between the output terminals a, b, c and the negative terminal of the dc bus as follows:

$v_{aN} = +V_{dc}$, when S_1 is ON, S_2 is OFF
$v_{aN} = 0$, when S_1 is OFF, S_2 is ON

Table 5.1 Three-phase voltages corresponding to switching combinations

	S_1	S_3	S_5	v_{an}	v_{bn}	v_{cn}	v_{ab}	v_{bc}	v_{ca}
$\overline{V}_0$	OFF	OFF	OFF	0	0	0	0	0	0
$\overline{V}_1$	ON	OFF	OFF	$\frac{2}{3}V_{dc}$	$-\frac{1}{3}V_{dc}$	$-\frac{1}{3}V_{dc}$	V_{dc}	0	$-V_{dc}$
$\overline{V}_2$	OFF	ON	OFF	$-\frac{1}{3}V_{dc}$	$\frac{2}{3}V_{dc}$	$-\frac{1}{3}V_{dc}$	$-V_{dc}$	V_{dc}	0
$\overline{V}_3$	OFF	OFF	ON	$-\frac{1}{3}V_{dc}$	$-\frac{1}{3}V_{dc}$	$\frac{2}{3}V_{dc}$	0	$-V_{dc}$	V_{dc}
$\overline{V}_4$	OFF	ON	ON	$-\frac{2}{3}V_{dc}$	$\frac{1}{3}V_{dc}$	$\frac{1}{3}V_{dc}$	$-V_{dc}$	0	V_{dc}
$\overline{V}_5$	ON	OFF	ON	$\frac{1}{3}V_{dc}$	$-\frac{2}{3}V_{dc}$	$\frac{1}{3}V_{dc}$	V_{dc}	$-V_{dc}$	0
$\overline{V}_6$	ON	ON	OFF	$\frac{1}{3}V_{dc}$	$\frac{1}{3}V_{dc}$	$-\frac{2}{3}V_{dc}$	0	V_{dc}	$-V_{dc}$
$\overline{V}_7$	ON	ON	ON	0	0	0	0	0	0

$v_{bN} = +V_{dc}$, when S_3 is ON, S_4 is OFF
$v_{bN} = 0$, when S_3 is OFF, S_4 is ON
$v_{cN} = +V_{dc}$, when S_5 is ON, S_6 is OFF
$v_{cN} = 0$, when S_5 is OFF, S_6 is ON

Using the above voltages v_{aN}, v_{bN}, v_{cN}, we can define both the line-line output voltages of the converter and the line-neutral voltages of the converter. The line-line output voltages are as follows:

$$v_{ab} = v_{aN} - v_{bN}$$
$$v_{bc} = v_{bN} - v_{cN} \tag{5.19}$$
$$v_{ca} = v_{cN} - v_{aN}$$

The line-neutral voltages on the other hand need some substitutions and rearrangement which has been discussed in a previous section. The voltages can be expressed as follows:

$$v_{an} = \frac{2}{3}v_{aN} - \frac{1}{3}v_{bN} - \frac{1}{3}v_{cN}$$
$$v_{bn} = -\frac{1}{3}v_{aN} + \frac{2}{3}v_{bN} - \frac{1}{3}v_{cN} \tag{5.20}$$
$$v_{cn} = -\frac{1}{3}v_{aN} - \frac{1}{3}v_{bN} + \frac{2}{3}v_{cN}$$

Using the above equations, one can compute line-line and line-neutral voltages for all the possible switching combinations in Fig. 5.5. Table. 5.1 lists the combinations with both line-line and line-neutral output voltages produced. Each switching combination has been designated as $\overline{V}_x$ ($\overline{V}_0$ to $\overline{V}_7$) as the output voltages produced by these switching combinations can be transformed into vectors using Clarke's transformation. In Table. 5.1, only the upper controllable devices S_1, S_3, S_5 have been listed as the lower controllable devices have conduction states that are complementary to the upper devices.

Table 5.2 Transformed voltages corresponding to switching combinations

	S_1	S_3	S_5	Line-neutral voltages		Line-line voltages	
				v_α	v_β	v_α	v_β
$\overline{V}_0$	OFF	OFF	OFF	0	0	0	0
$\overline{V}_1$	ON	OFF	OFF	$0.6667V_{dc}$	0	V_{dc}	$0.5773V_{dc}$
$\overline{V}_2$	OFF	ON	OFF	$-0.3333V_{dc}$	$0.5773V_{dc}$	$-V_{dc}$	$0.5773V_{dc}$
$\overline{V}_3$	OFF	OFF	ON	$-0.3333V_{dc}$	$-0.5773V_{dc}$	0	$-1.1547V_{dc}$
$\overline{V}_4$	OFF	ON	ON	$-0.6667V_{dc}$	0	$-V_{dc}$	$-0.5773V_{dc}$
$\overline{V}_5$	ON	OFF	ON	$0.3333V_{dc}$	$-0.5773V_{dc}$	V_{dc}	$-0.5773V_{dc}$
$\overline{V}_6$	ON	ON	OFF	$0.3333V_{dc}$	$0.5773V_{dc}$	0	$1.1547V_{dc}$
$\overline{V}_7$	ON	ON	ON	0	0	0	0

The reader is encouraged to perform calculations for each switching combination in Table 5.1 and verify that the line-neutral and line-line voltages for each switching combination are as listed in Table 5.1 [8]. From Table. 5.1, we can extract two sets of three-phase quantities v_{an}, v_{bn}, v_{cn} and v_{ab}, v_{bc}, v_{ca}. For each switching combination, it can be observed that the sum of the three-phase quantities are always zero, namely:

$$v_{an} + v_{bn} + v_{cn} = 0$$
$$v_{ab} + v_{bc} + v_{ca} = 0$$

(5.21)

As a result of the above relations holding true for every possible switching combination, performing Clarke's transformation on the line-neutral and line-line output voltages will always result in the 0 component being zero and therefore, only the α and β components need to be calculated. Table 5.2 lists v_α and v_β for the line-neutral and line-line voltages of Table 5.1. The reader is encouraged to use Clarke's transformation on the values in Table 5.1 and verify the values in Table 5.2. Besides the zero vectors $\overline{V}_0$ and $\overline{V}_7$, the remaining vectors $\overline{V}_1$ to $\overline{V}_6$ have non-zero values and can be drawn on the α-β complex plane. Figures 5.12 and 5.13 show the vector diagrams for line-neutral voltages and line-line voltages respectively.

In the vector diagrams of Figs. 5.12 and 5.13, the vectors are plotted on the complex α-β plane assuming $V_{dc} = 1$ as one can substitute a practical value of V_{dc} and obtain the exact voltage vectors. Next to each vector, a set of coordinates (S_a, S_b, S_c) can be found which indicates the switching combination to which the vector corresponds, with $S_x = 1$ indicating that an upper device in the leg is conducting, while $S_x = 0$ indicating that a lower device in the leg is conducting. In both vector diagrams as well as in Table 5.2 it can be observed that all the voltage vectors have the same absolute magnitude $\sqrt{v_\alpha^2 + v_\beta^2}$, with the line-neutral voltage vectors having an absolute magnitude of $0.6667V_{dc}$ and the line-line voltage vectors having an absolute magnitude of $1.1547V_{dc}$ [9]. The voltage vectors of Figs. 5.12 and 5.13 have been produced by a Python program that the reader can find in the

Fig. 5.12 Vectors in the complex α-β plane for line-neutral voltages

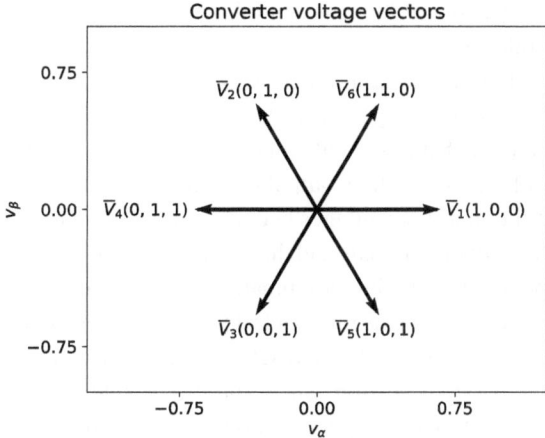

Fig. 5.13 Vectors in the complex α-β plane for line-line voltages

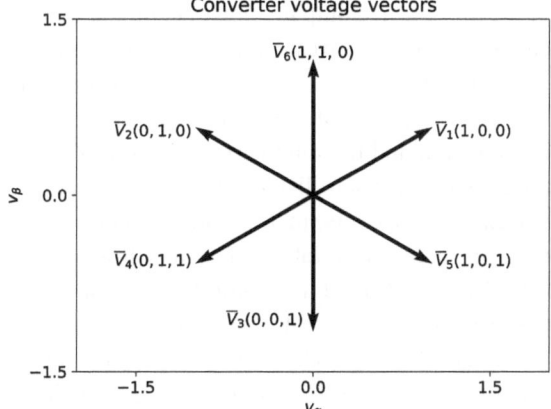

folder `vector_diagrams` within the folder `chapter5` at the link: https://github.com/opensourceelectrical/switching-strategies-for-power-electronics.

These programs will also be used in the next section while formulating a switching strategy for the three-leg converter.

When one compares the vector diagrams of Figs. 5.12 and 5.13 with the vector diagram of the full-bridge converter in the previous chapter, a few very interesting observations can be made. It appears as if in both the three-leg converter and the full-bridge converter, the vectors span the 360° two dimensional space in an equidistant manner with the angle between vectors equal to $\frac{360°}{N}$ with N being the number of non-zero vectors. Since the full-bridge converter had only two non-zero vectors, the vectors are 180° apart. In the case of the three-phase converter, with six non-zero vectors, the vectors form the corners of a hexagon and are 60° apart from one another. The hexagons of Figs. 5.12 and 5.13 differ in both the magnitude of vectors as well as the position of the vectors, as one of them contains the vectors of the line-neutral voltages while the other contains the vectors of the line-line voltages.

Let us take a moment to summarize the process above. For the three-leg three-phase converter of Fig. 5.4, we have already established in a previous section that the converter can have 8 unique switching combinations as shown in Fig. 5.5. Since the three-leg converter of Fig. 5.4 has three output terminals, we can express the output voltages of the converter as either line-line voltages between these output terminals, or as line-neutral voltages between the output terminals and the neutral of a star connected load or machine. Therefore, for every possible switching combination, we can calculate the output voltages of the converter, as either line-line or line-neutral voltages. These line-line or line-neutral three-phase voltages can be transformed using Clarke's transformation to the α-β complex plane as described in the previous section. Since for all possible switching combinations, the sum of the three-phase output voltages are zero in both line-line and line-neutral cases, the α-β complex plane completely represents the converter three-phase output voltages. With α and β components of the transformed output voltages computed for every possible switching combination, we can have a vector corresponding to each switching combination in the complex α-β plane.

If we examine the vectors and switching combinations in Table. 5.2 and compare them to the vector diagrams of Figs. 5.12 and 5.13, we can form a correspondence with a switching combination and the output voltage produced. As an example, the vector $\overline{V}_1$ that is produced when the controllable device S_1 conducts (or diode D_1 depending on the direction of current), S_4 conducts (or D_4) and S_6 conducts (or D_6) will have line-neutral and line-line voltages as listed in Table 5.1, and will have a complex value in the α-β plane equal to $0.6667V_{dc} + j0$ for line-neutral voltages and $(1 + j0.5773)V_{dc}$ for line-line voltages. In comparison, the vector $\overline{V}_3$ (S_2−ON, S_4−ON, S_5−ON) will produce in the α-β plane a value equal to $(-0.3333 - j0.5773)V_{dc}$ for line-neutral voltages and $0 - j1.4142V_{dc}$ for line-line voltages. This mapping between Table 5.2 and the vector diagrams of Figs. 5.12 and 5.13 is merely mathematical, but to fully understand the significance of how these vector diagrams can be used to control a converter, one needs to go back to rotating vectors.

If we have a balanced set of three-phase voltages, we can represent them as rotating vectors as shown in Fig. 5.3. Performing Clarke's transformation of these rotating vectors as shown in Fig. 5.9 produces the complex plane α-β representation of Fig. 5.10. Due to the sinusoidal nature of the components in the α-β complex plane, the vector that is formed from the α and β component also resembles a rotating vector as time changes. Therefore, as the three-phase vectors rotate at an angular speed ω in the counter-clockwise direction, the vector in the complex α-β plane also rotates at an angular speed ω in the counter-clockwise direction. For a particular instant of time, for a given position of the three-phase vectors, there will be v_α and v_β which will have unique instantaneous values which will be the projections of the vectors on the horizontal axis. These instantaneous values of v_α and v_β in the complex α-β plane will result in a unique vector $\overline{V}$. Therefore, there exists a unique mapping between the three-phase rotating vectors and the rotating vector $\overline{V}$ in the complex α-β plane.

When we control a three-leg converter such as the one in Fig. 5.4, our objective is that the converter should produce three-phase voltages that are as close as possible to a balanced three-phase sinusoidal set. Since the converter is a power electronic converter, it will not be able to produce sinusoidal voltages. But as in the previous chapters, the converter can be controlled to produce a switched voltage which contains these sinusoidal voltages besides also other harmonic components. To remove the harmonic components and leave only the sinusoidal components, we must design appropriate low pass filters. Therefore, if the converter is to produce output voltages that contain balanced sinusoids, the vector $\overline{V}$ that results from performing Clarke's transformation on the converter output voltages must be a rotating vector that rotates in the counter-clockwise direction with an angular speed ω. From Figs. 5.12 and 5.13, the transformed voltage vector $\overline{V}$ must rotate smoothly from $\overline{V}_1$ to $\overline{V}_6$ to $\overline{V}_2$ to $\overline{V}_4$ to $\overline{V}_3$ to $\overline{V}_5$ and back to $\overline{V}_1$. Since we have randomly chosen the vectors with respect to the switching combinations, the vectors do not transition in a numerical order. However, this is merely a matter of choice, and if we so wish, we can always rearrange the switching combinations in Tables 5.1 and 5.2 to produce vector transitions that change from 1 to 6 and back to 1.

This, therefore, is the challenge in designing a switching strategy for the three-leg converter. To be able to choose switching combinations such that the resultant output vector rotates smoothly in the counter-clockwise sense at the same angular speed as that of the grid. A reader can find vast literature that deals with this topic which has been actively researched for the past few decades. However, for a newcomer to power electronics, attempting to understand these switching strategies can be fairly difficult. In this section, we have created a vector representation of the three-leg converter based on our examination of Clarke's transformation in the previous section. In the next section, we will use this vector hexagon of the converter to describe a very popular modulation strategy called as Space Vector Pulse Width Modulation (SVPWM).

5.6   Space Vector Pulse Width Modulation (SVPWM)

In the previous section, we examined how using Clarke's transformation, the three-phase voltages produced by the three-leg converter of Fig. 5.4 can be represented as a vector hexagon with one unique vector corresponding to every possible switching combination. In this section, we will examine how we can superimpose the desired three-phase output voltages onto this vector diagram, and thereafter, determine the converter voltage vectors that can be chosen to produce the required output voltage. This section will deal with algorithms and strategies and will present simulations to describe the implementation of the switching strategy. The reader is encouraged to review the previous sections on Clarke's transformation and on the vector representation of a three-leg converter.

The three-leg converter of Fig. 5.4 can feed a three-phase load or machine or could be interfaced to a three-phase distribution system. The load, machine or distribution system could be connected in any number of ways as already described in the previous section. However, for simplicity, let us consider only the star and delta connections. For a load, machine or distribution system connected in delta, one can only define the line-line voltages between the a, b, c terminals. In such a case, we can superimpose the result of the Clarke's transformation of the desired three-phase line-line voltages on the line-line converter voltage vector diagram of Fig. 5.13. If however, the load, machine or distribution system were connected in star with a neutral terminal, we can define either line-line voltages or line-neutral voltages. Therefore, the required three-phase output voltages of the converter can be expressed either between the lines and the neutral or between the lines. The result of the Clarke's transformation of the line-neutral three-phase output voltages can be superimposed on the line-neutral converter voltage vector diagram of Fig. 5.12. Therefore, however we wish to define the three-phase output voltages that we expect the converter to produce, we have a corresponding converter voltage vector diagram.

Let us elaborate on this with an example. Since in the vector diagrams of Figs. 5.12 and 5.13, we had assumed $V_{dc} = 1$, let us suppose that we wish to produce an attenuated set of three-phase line-neutral voltages as follows:

$$
\begin{aligned}
v_{ran} &= 0.3 \cos(\omega t) \\
v_{rbn} &= 0.3 \cos(\omega t - 120°) \\
v_{rcn} &= 0.3 \cos(\omega t - 240°)
\end{aligned}
\tag{5.22}
$$

where the subscript r stands for the required value. On performing Clarke's transformation, the resultant voltages will be:

$$
\begin{aligned}
v_{r\alpha} &= 0.3 \cos(\omega t) \\
v_{r\beta} &= 0.3 \sin(\omega t)
\end{aligned}
\tag{5.23}
$$

The transformed voltage can be expressed in the complex α-β plane as another rotating vector:

$$
\overline{V}_r = v_{r\alpha} + j v_{r\beta} = 0.3[\cos(\omega t) + j \sin(\omega t)]
\tag{5.24}
$$

Figure 5.14 shows the line-neutral converter voltage vector diagram with the transformed required three-phase ac voltages superimposed as a vector having a circular trajectory. The required voltage vector will rotate in the counter-clockwise direction and will assume different positions with the tip on the dotted circle. As an example, for a certain time instant t_1, if $\omega t_1 = 30°$, $\overline{V}_{r1} = 0.26 + j0.15$, while at another time instant t_2, if $\omega t_2 = 260°$, $\overline{V}_{r2} = -0.052 - j0.295$. These two vectors are depicted in Fig. 5.14. Therefore, at any given instant of time, since the three-phase line-neutral ac voltages are known, one can calculate the transformed voltages in the α-β complex plane, and plot the exact position of the vector along with the converter line-neutral voltage vectors. Though this might seem to be fairly clear and obvious, to make this description complete, a few details about closed-loop control will be of great help.

Fig. 5.14 Vectors in the complex α-β plane for line-neutral voltages

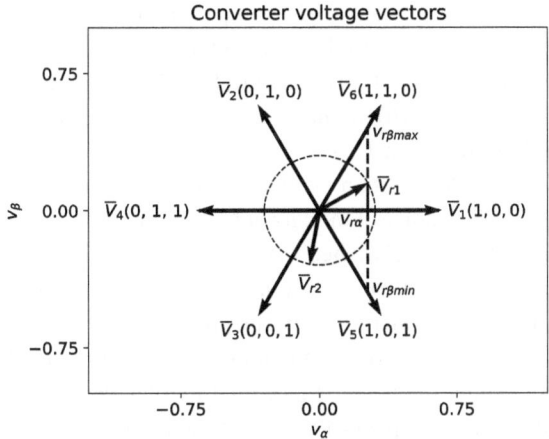

In the previous chapters and also in the first simulation in this chapter, the modulation signals (or duty ratios) were the inputs provided to generate switching signals for the converters. These modulation signals were generated by a closed-loop control that may consist of one or more loops that condition one or more system quantities such as output voltages or output currents. Let us now consider the three-leg converter supplying either a load, machine or a distribution system. The final system quantities that need to be regulated will depend on the application. If the converter were supplying a load or a machine, the requirement is usually that the voltages applied across the load or machine terminals are regulated to certain desired values. Either these voltages will need to be merely smooth balanced sinusoids of a constant magnitude or they will need to be such that the machine performs as desires i.e it accelerates, decelerates or produces a certain torque. If the converter were supplying a distribution system, the requirement is usually to supply currents that support the distribution system in a certain manner i.e reactive power compensation, harmonic compensation etc. Though the final requirement is to supply currents, the converter achieves this by producing certain output voltages with respect to the distribution system voltages.

Therefore, from a control perspective, quite often, the immediate variables from which switching signals can be generated are the output voltages that the converter needs to produce. With modifications to the control scheme, it is always possible to translate these output voltages to modulation signals with which sine-triangle PWM can be implemented. However, if we were to generate switching signals directly from the output voltages that the converter is required to generate, we need to find a way to ensure that the converter is able to do so. From Table 5.2 and the vector diagrams of Figs. 5.12 and 5.13, the three-leg converter can only produce six output voltages corresponding to each switching combination with two additional zero vectors. Therefore, to define the problem that we are trying to solve, we need a switching strategy whereby we can generate a smooth rotating output voltage vector shown by the dotted circle in Fig. 5.14 using only the six vectors that the converter can produce. As before, the fundamental concept behind generating switching signals for a converter lies in generating a pattern by using a number of switching combinations over a switching cycle.

In the case of sine-triangle PWM, we defined a carrier waveform as either a sawtooth waveform or a triangular waveform with a frequency equal to the desired switching frequency of the converter. This carrier waveform established the heart beat of the converter, and within each cycle a certain switching pattern was generated similar to repeated pattern of a heart beat. The choice of the carrier being a sawtooth waveform or a triangular waveform depended on whether we wished the switching pattern to be symmetrical over a switching cycle or not, as already described in detail over the previous chapters. If we do not use modulation signals, and therefore, do not need carrier signals for comparison, we will still need to ensure that a certain switching pattern is produced in a time interval corresponding to the switching frequency. Since the three-leg converter produces ac voltages, we would require the switching pattern to be symmetrical over the switching cycle. Therefore, the first step is to choose a switching frequency of the converter, and over each switching cycle to use multiple switching combinations such that average time-weighted voltage produced is equal to the desired output voltage in the switching cycle.

In these modern times, controllers are implemented digitally using either DSP microcontrollers or FPGAs, and therefore, the control algorithms are executed at a fixed time interval. This control time interval may be the same as the switching cycle time interval or could be different. In case the two time intervals are different, the control time interval is usually greater than the switching time interval, as in some cases closed-loop control algorithms can be fairly complex and need time to be processed by the microcontroller. Therefore, a particular three-phase output voltage produced by the control algorithm can be available either for a single switching cycle or for several switching cycles. Since the switching frequency is usually much greater than the grid frequency, the output voltages can be assumed to be fixed over a switching cycle. Microcontrollers can be configured in various ways to execute segments of code at various time intervals and using various synchronization schemes. However, in most cases, it is fairly safe to assume that at the beginning of a switching cycle, three-phase output voltages are available from the control algorithm.

If we return to the vector diagram of Fig. 5.14, we have drawn a trajectory of the required output voltage as a circle. There will be specific time instants when the required output voltage vector will be coincident with one of the six voltage vectors of the converter. In such a case, over a switching cycle, one only needs to apply the following relation:

$$\overline{V}_r T = \overline{V}_x t_1 + \overline{V}_0 t_0 = \overline{V}_x t_1 \tag{5.25}$$

In the above expression, T is the time interval of the switching cycle, $\overline{V}_x$ is one of the six converter voltage vectors, $\overline{V}_0$ is the zero output voltage vector, t_1 and t_0 are the time intervals of the non-zero and zero voltage vectors respectively. Simply speaking, since the instantaneous values of the required output voltage $\overline{V}_r$ is less than the converter voltage $\overline{V}_x$, it is necessary to add a zero vector for a finite time duration to decrease the average output voltage produced. Furthermore, since the above expression is a complex equation, one can separate them into real and imaginary parts as follows:

$$v_{r\alpha}T = v_{x\alpha}t_1$$
$$v_{r\beta}T = v_{x\beta}t_1$$

(5.26)

It is obvious that since we only wish to calculate time period t_1, there is no need for two equations, and we can use any one of the equations to calculate t_1. Once t_1 is obtained, the time interval for the zero voltage vector is merely:

$$t_0 = T - t_1$$

(5.27)

In the above expression, we have written the zero vector to be $\overline{V}_0$. However, we have two zero voltage vectors $\overline{V}_0$ and $\overline{V}_7$ as seen in Tables 5.1 and 5.2. The question arises, which one should be chosen? The answer depends on the non-zero voltage vector $\overline{V}_x$, as the zero vector should be chosen to ensure minimum switching to decrease switching losses of the converter. For example, if the non-zero voltage vector was $\overline{V}_1(1, 0, 0)$, the zero voltage vector would be $\overline{V}_0(0, 0, 0)$, as in that case, only the devices in one leg (a) change conduction state, namely controllable device S_1 (or D_1) will stop conducting and device S_2 (or D_2) will start conducting. On the other hand, if the non-zero voltage vector was $\overline{V}_5(1, 0, 1)$, the zero voltage vector should be $\overline{V}_7(1, 1, 1)$, as this will again result in the devices in only one leg (b) changing conduction states.

Though it might seem like a solution to use one non-zero vector and one zero vector for these conditions when the required output voltage vector is coincident with one of the six converter voltage vectors, it is important to note that these occurrences are rare. From mere visual inspection of Fig. 5.14, as the required output voltage vector rotates around the dotted circle, the probability that at a particular instant of time when the switching cycle begins, the required output voltage vector will coincide with a converter voltage vector is very low. It is much more likely that the required output voltage will lie between two converter voltage vectors, and two examples of these are shown in Fig. 5.14. In such a case, to produce the required output voltage, we cannot merely choose one of the non-zero converter voltage vectors and a zero vector. We must choose both converter voltage vectors between which the required output voltage vector lies in addition to zero vectors. Therefore, if the required output voltage vector lies between two non-zero converter voltage vectors $\overline{V}_x$ and $\overline{V}_y$, the expression becomes:

$$\overline{V}_r T = \overline{V}_x t_1 + \overline{V}_y t_2$$

(5.28)

Before we compute the time periods t_1 and t_2, we must first determine the converter voltage vectors $\overline{V}_x$ and $\overline{V}_y$. Since the converter voltage vectors form the corners of a hexagon, there are six possible combinations of $\overline{V}_x, \overline{V}_y$ that are possible based on the precise location of the required output voltage vector $\overline{V}_r$. From both Figs. 5.12 and 5.13, these combinations can be observed to be $(\overline{V}_1, \overline{V}_6), (\overline{V}_6, \overline{V}_2), (\overline{V}_2, \overline{V}_4), (\overline{V}_4, \overline{V}_3), (\overline{V}_3, \overline{V}_5)$ and $(\overline{V}_5, \overline{V}_1)$. There are many algorithms proposed to determine the converter voltage vectors between which the required voltage vector lies. One method is to calculate the angle that the required voltage vector makes with the α axis using the expression:

$$\theta = \tan^{-1} \frac{v_{r\beta}}{v_{r\alpha}} \tag{5.29}$$

Once this angle is known, using a few conditional checks, it is possible to determine the converter voltage vectors $\overline{V}_x$, $\overline{V}_y$ as the angles made by the converter voltage vectors with the α axis can be calculated and depend only on whether the voltages are line-neutral or line-line.

Though calculation of the angle θ might seem fairly simple, there are a few cases in a hardware implementation that might cause issues. To begin with, the computation of the inverse tangent is now possible in advanced floating point microprocessors, as they have in-built mathematical functions that are processed rapidly. However, for basic microcontrollers, the computation of the inverse tangent might be an expensive computation that might require several clock cycles. In addition, when the fraction $\frac{v_{r\beta}}{v_{r\alpha}}$ is computed, this number might become very large when $v_{r\alpha}$ is close to zero. This will occur when the angle θ approaches $90°$ or multiples of $90°$. Large numbers might result in overflows depending on the precision of the microprocessor, which in turn can result in unpredictable behaviour. For these reasons, in this book, angle calculation using the above expression will not be performed for determining the converter voltage vectors. However, if readers do wish to use angle calculation as an implementation, they should check with the capabilities of the microcontroller that they are using.

Since the components $v_{r\alpha}$, $v_{r\beta}$ of the required output voltage vector are known, one can implement a fairly simple algorithm using these components and the knowledge of the converter voltage vectors. The algorithm will differ for line-neutral voltage vectors of Fig. 5.12 and for line-line voltage vectors of Fig. 5.13. Let us describe in detail the algorithm for the line-neutral voltage vectors of Fig. 5.12. When $v_{r\alpha} > 0$, the required output voltage vector can lie between the converter voltage combinations $(\overline{V}_1, \overline{V}_6)$, $(\overline{V}_6, \overline{V}_2)$, $(\overline{V}_3, \overline{V}_5)$ and $(\overline{V}_5, \overline{V}_1)$. For $v_{r\beta} > 0$, the combinations are further reduced to $(\overline{V}_1, \overline{V}_6)$ and $(\overline{V}_6, \overline{V}_2)$, while for $v_{r\beta} < 0$, the combinations are $(\overline{V}_3, \overline{V}_5)$ and $(\overline{V}_5, \overline{V}_1)$. To determine the exact combination, one needs to use the converter voltage vectors $\overline{V}_6$ and $\overline{V}_5$ and the fact that their angle with respect to the α axis is $60°$ and $-60°$ respectively. For a particular value $v_{r\alpha}$, one can define these two limits:

$$\begin{aligned} v_{r\beta max} &= v_{r\alpha} \tan 60° \\ v_{r\beta min} &= -v_{r\alpha} \tan 60° \end{aligned} \tag{5.30}$$

$v_{r\beta max}$ and $v_{r\beta min}$ are indicated in Fig. 5.14 and can be seen to be values of $v_{r\beta}$ when the required output voltage vector lies on $\overline{V}_6$ and $\overline{V}_5$ respectively. Therefore, if $v_{r\beta} > 0$ and $v_{r\beta} < v_{r\beta max}$, the required output voltage vector lies between $(\overline{V}_1, \overline{V}_6)$. If $v_{r\beta} > 0$ but $v_{r\beta} > v_{r\beta max}$, the required output voltage vector lies between $(\overline{V}_6, \overline{V}_2)$. If $v_{r\beta} < 0$ and $v_{r\beta} > v_{r\beta min}$, the required output voltage vector lies between $(\overline{V}_5, \overline{V}_1)$. If $v_{r\beta} < 0$ but

$v_{r\beta} < v_{r\beta min}$, the required output voltage vector lies between $(\overline{V}_3, \overline{V}_5)$. This algorithm has the significant advantage that it needs only simple computations. The value $\tan 60° = 1.732$ is a constant and can be introduced as a constant rather than performing the computation in a microcontroller. The algorithm also does not suffer from any stability issues when either $v_{r\alpha}$ or $v_{r\beta}$ are close to zero. The entire algorithm can be visualized by the flowchart of Fig. 5.15. As can be observed, a maximum of three condition checks are needed to determine the converter voltage vectors.

Once the converter voltage vectors have been identified, the next question is about how these should be applied during a switching cycle. We are aware that along with these two converter voltage vectors $(\overline{V}_x, \overline{V}_y)$, we would need to apply zero voltage vectors as well. Therefore, we need to select a sequence in which all these vectors can be applied. In order to select a sequence, one needs to take into account two factors. The first factor being that the vectors should be chosen in a such a manner that the minimum number of switching transitions occur in order to minimize switching losses. The second factor being that the vectors should be chosen in such a manner that the switching pattern is symmetrical over the switching cycle. In order to satisfy the first criterion, we need to ensure that the transition between the end of a switching cycle and the beginning of another switching cycle is as smooth as possible. This can be achieved by choosing the first and last vector in the sequence to be the zero vector $\overline{V}_0$. Therefore, when one switching cycle ends and another one begins, there is no change in the switching states of the converter. Moreover, this choice also maintains the symmetry of the switching pattern over the switching cycle.

To fully satisfy the first criterion, one needs to closely inspect the vector diagram of Fig. 5.14 and make an interesting observation that holds true for every pair of converter voltage vectors that are adjacent to each other. One will find that one of the converter voltage vectors differs from the zero vector $\overline{V}_0$ by a single switching state, while the other converter voltage vector differs from the zero vector $\overline{V}_7$ by a single switching state. As an example, for the vector combination $(\overline{V}_2, \overline{V}_4)$, the vector $\overline{V}_2(0,1,0)$ differs from the zero vector $\overline{V}_0(0,0,0)$ by only one switching transition, namely device S_3 (or D_3) in leg b starting to conduct and S_4 (or D_4) stopping conduction. The vector $\overline{V}_4(0,1,1)$ differs from the zero vector $\overline{V}_7(1,1,1)$ by a single switching transition, namely device S_1 (or D_1) in leg a starting to conduct and S_2 (or D_2) stopping conduction. Furthermore, the vectors $\overline{V}_2$ and $\overline{V}_4$ themselves differ only by single switching transition, namely device S_5 (or D_5) in leg c starting to conduct and S_6 (or D_6) stopping conduction. The reader is encouraged to verify that this holds true for all six vector combinations that have been listed so far and can be observed from the vector diagram of Fig. 5.14 [10].

This might appear to be similar to a clue in a puzzle where the pieces conveniently come together once one figures it out. However, this property is merely the result of the switching combinations producing certain three-phase voltages, which subsequent to Clarke's transformation end up in a sequence. When listing the possible switching combinations in Fig. 5.5, the procedure used was simple binary permutation. However, in reality, the sequence of voltage vectors corresponding to the switching combinations are aligned such that a sin-

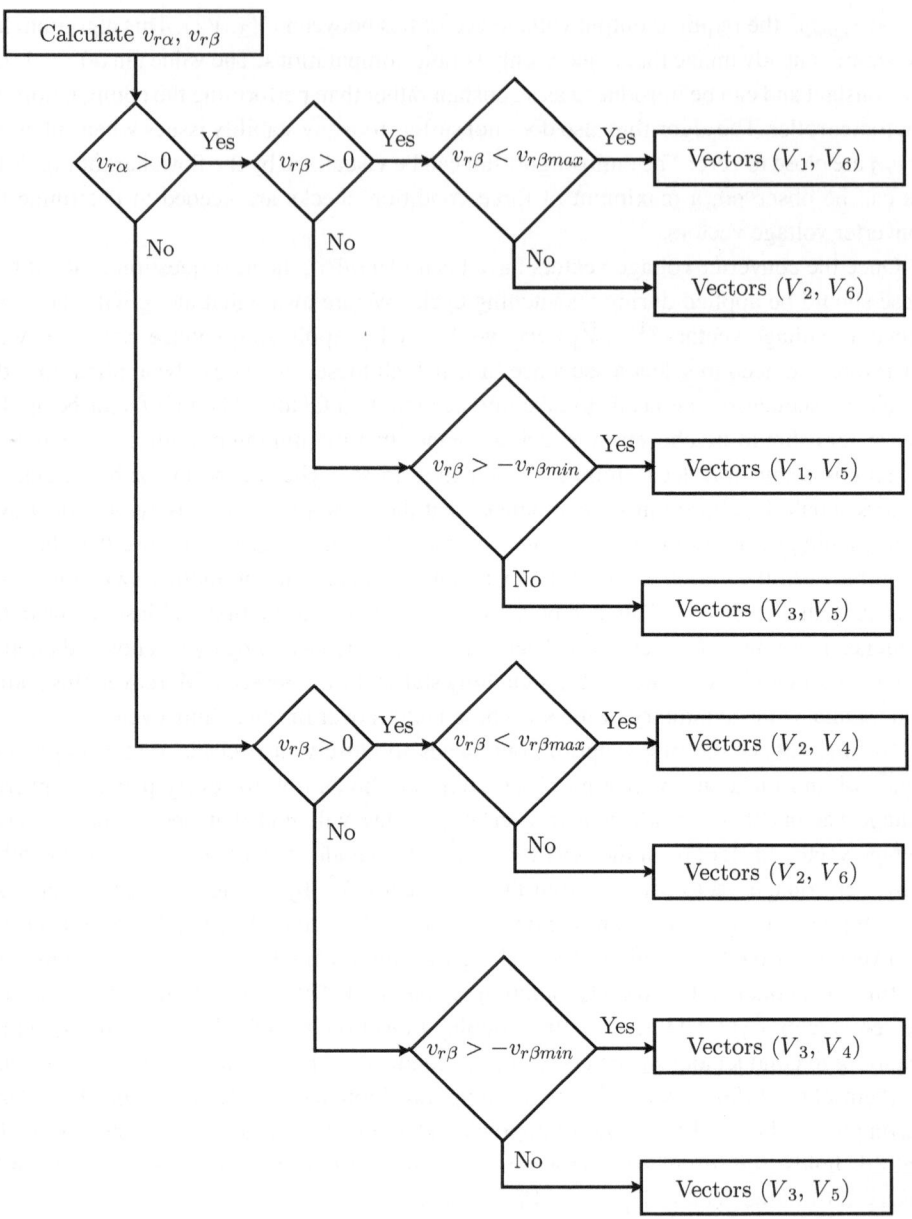

Fig. 5.15 Algorithm for determining converter voltage vectors

gle switching transition in one leg results in voltage vectors being adjacent to each other. The result of this observation is that we can now line up the converter voltage vectors as follows—the zero vector $\overline{V}_0$, followed by one of the non-zero voltage vectors that differs from $\overline{V}_0$ by a single switching transition, the other non-zero voltage vector and finally the other zero voltage vector $\overline{V}_7$. This line-up does satisfy one of the criteria, which is to minimize the switching transitions. However, this line-up violates the second criterion, which is to produce a symmetrical switching pattern over the switching cycle.

To satisfy the second criterion, we need to perform an adaptation. If we return back to the expression of (5.28), where t_0 is the time period for which the zero vector is applied, and t_1 and t_2 are the time periods for which the non-zero vectors are applied, the sequence of voltage vectors above would be—$\overline{V}_0$ for $\frac{t_0}{2}$ interval, $\overline{V}_x$ for t_1 interval, $\overline{V}_y$ for t_2 interval and $\overline{V}_7$ for $\frac{t_0}{2}$ interval. We have merely equally divided by the time period t_0 into two parts for applying the zero vectors $\overline{V}_0$ and $\overline{V}_7$. As long as we maintain the above proportions of the voltage vectors, we are free to make adjustments as we wish. Therefore, if we further divide the time period of all four vectors by 2, we can create a mirror image of the vector sequence and satisfy the second criterion. The sequence becomes $\overline{V}_0$ for $\frac{t_0}{4}$, $\overline{V}_x$ for $\frac{t_1}{2}$, $\overline{V}_y$ for $\frac{t_2}{2}$, $\overline{V}_7$ for $\frac{t_0}{4}$, $\overline{V}_7$ for $\frac{t_0}{4}$, $\overline{V}_y$ for $\frac{t_2}{2}$, $\overline{V}_x$ for $\frac{t_1}{2}$ and $\overline{V}_0$ for $\frac{t_0}{4}$. It is easy to verify that the time periods of the zero and non-zero vectors are unchanged, and the sequence is now symmetrical over the switching cycle.

With the above sequence, only one issue remains, which is the selection of the non-zero vectors $\overline{V}_x$ and $\overline{V}_y$. From an observation of Fig. 5.14, it can be seen that if one progresses in the counter-clockwise direction and inspects every pair of voltage vectors, the selection of $\overline{V}_x$ and $\overline{V}_y$ will change. As an example, with the vector combination $(\overline{V}_1, \overline{V}_6)$, the voltage vector $\overline{V}_1(1,0,0)$ is closer to the zero vector $\overline{V}_0(0,0,0)$ as compared to the voltage vector $\overline{V}_6(1,1,0)$. On the other hand, for the next vector combination $(\overline{V}_6, \overline{V}_2)$ in the counter-clockwise direction, the vector $\overline{V}_2(0,1,0)$ is closer to the zero vector $\overline{V}_0(0,0,0)$ as compared to $\overline{V}_6(1,1,0)$. There are several ways to sort this interchange. The first and simplest method is to define a sequence table for every combination of vectors, and store this table in a convenient location in the microcontroller memory. Since there are only six possible combinations and four vectors per combination, the memory requirements are not prohibitive, and even basic microcontrollers will have sufficient memory to store this data. Upon determining that a particular vector combination is to be applied, the entire sequence can be fetched from memory. Such a sequence table is listed in Table 5.3.

If one does not wish to generate an entire sequence table such as in Table 5.3, there are a few other techniques to compute which non-zero vector should be chosen first. One method is use the sum of the ON states in a vector as an indication to how close it is to a given zero voltage vector. For example, for the vector combination $(\overline{V}_6, \overline{V}_2)$, the voltage vector $\overline{V}_6(1,1,0)$ has a ON state sum $1 + 1 + 0 = 2$ while the voltage vector $\overline{V}_2(0,1,0)$ has a ON state sum $0 + 1 + 0 = 1$. Therefore, with a lower number of upper devices conducting, the voltage vector $\overline{V}_2$ is closer to the zero vector $\overline{V}_0$, while with a larger number of upper devices conducting, the voltage vector $\overline{V}_6$ is closer to the zero vector $\overline{V}_7$. All we need to

Table 5.3 Sequence of converter voltage vectors for different pairs of converter voltage vectors

	$\frac{t_0}{4}$	$\frac{t_1}{2}$	$\frac{t_2}{2}$	$\frac{t_0}{4}$	$\frac{t_0}{4}$	$\frac{t_2}{2}$	$\frac{t_1}{2}$	$\frac{t_0}{4}$
$(\overline{V}_1,\overline{V}_6)$	$\overline{V}_0$	$\overline{V}_1$	$\overline{V}_6$	$\overline{V}_7$	$\overline{V}_7$	$\overline{V}_6$	$\overline{V}_1$	$\overline{V}_0$
$(\overline{V}_6,\overline{V}_2)$	$\overline{V}_0$	$\overline{V}_2$	$\overline{V}_6$	$\overline{V}_7$	$\overline{V}_7$	$\overline{V}_6$	$\overline{V}_2$	$\overline{V}_0$
$(\overline{V}_2,\overline{V}_4)$	$\overline{V}_0$	$\overline{V}_2$	$\overline{V}_4$	$\overline{V}_7$	$\overline{V}_7$	$\overline{V}_4$	$\overline{V}_2$	$\overline{V}_0$
$(\overline{V}_4,\overline{V}_3)$	$\overline{V}_0$	$\overline{V}_3$	$\overline{V}_4$	$\overline{V}_7$	$\overline{V}_7$	$\overline{V}_4$	$\overline{V}_3$	$\overline{V}_0$
$(\overline{V}_3,\overline{V}_5)$	$\overline{V}_0$	$\overline{V}_3$	$\overline{V}_5$	$\overline{V}_7$	$\overline{V}_7$	$\overline{V}_5$	$\overline{V}_3$	$\overline{V}_0$
$(\overline{V}_5,\overline{V}_1)$	$\overline{V}_0$	$\overline{V}_1$	$\overline{V}_5$	$\overline{V}_7$	$\overline{V}_7$	$\overline{V}_5$	$\overline{V}_1$	$\overline{V}_0$

store is the switching combination corresponding to every voltage vector which will occupy much lesser memory than the table of sequence vectors of Table 5.3. Another strategy can be formulated through observation of the vector diagram of Fig. 5.14. The vector combinations can be given sector identifiers with the combination $(\overline{V}_1, \overline{V}_6)$ forming sector 1, $(\overline{V}_6, \overline{V}_2)$ forming sector 2 and onwards until the last combination in the counter-clockwise sense $(\overline{V}_5, \overline{V}_1)$ forming sector 6. In the case of the odd numbered sectors, the first vector in the combination is the first in the sequence, while in the case of even numbered sectors, the second vector in the combination is the first in the sequence. In this case, all that needs to be stored is the switching combinations of the voltage vectors in sequence as we progress in the counter-clockwise direction.

Now that the vector combinations have been identified, we can return to solving expression (5.28) for the time intervals t_1 and t_2. The expression of (5.28) can be separated into real and imaginary parts:

$$v_{r\alpha}T = v_{x\alpha}t_1 + v_{y\alpha}t_2$$
$$v_{r\beta}T = v_{x\beta}t_1 + v_{y\beta}t_2$$

(5.31)

The above simultaneous equations can be solved by simple substitution:

$$t_1 = \frac{v_{r\alpha}v_{y\beta} - v_{y\alpha}v_{r\beta}}{v_{x\alpha}v_{y\beta} - v_{x\beta}v_{y\alpha}} T$$
$$t_2 = \frac{v_{r\beta}v_{x\alpha} - v_{x\beta}v_{r\alpha}}{v_{x\alpha}v_{y\beta} - v_{x\beta}v_{y\alpha}} T$$

(5.32)

Following the calculation of the time periods of the non-zero voltage vectors, the time period of the zero voltage vectors is:

$$t_0 = T - (t_1 + t_2)$$

(5.33)

Before closing the section, we should return to the special case of when the required output voltage vector is incident on one the converter voltage vectors. Since there are only six such cases, the reader is encouraged to apply the algorithm of Fig. 5.15 to determine which voltage vector combination will be selected, and thereafter use expression (5.32) to calculate time periods t_1 and t_2. As an example, if the required output voltage vector is incident along the converter voltage vector $\overline{V}_6$, the algorithm of Fig. 5.15 will select ($\overline{V}_6$, $\overline{V}_2$) as the vector combination. For this voltage vector combination, $\overline{V}_x = \overline{V}_2$ and $\overline{V}_y = \overline{V}_6$ from Table 5.3. Since the required output voltage vector $\overline{V}_r$ is incident on $\overline{V}_6$, the following expression can be written:

$$\overline{V}_r = k\overline{V}_6 = k\overline{V}_y \qquad (5.34)$$

where k is a number less than or equal to 1. And therefore:

$$\begin{aligned} v_{r\alpha} &= kv_{6\alpha} = kv_{y\alpha} \\ v_{r\beta} &= kv_{6\beta} = kv_{y\beta} \end{aligned} \qquad (5.35)$$

Using this relation in expression (5.32), the time period t_1 turns out to be zero and only t_2 is non-zero, which in turn implies that the zero vector time period t_0 will be accordingly adjusted. Therefore, for the special case of the required output voltage vector incident on a converter voltage vector, one obtains only a zero vector time period and a single non-zero vector time period. However, if one uses the same sequence as in Table 5.3, the transition from the zero vector $\overline{V}_0$ to the non-zero vector and then to the zero vector $\overline{V}_7$, this will result in excessive switching in almost all the six cases. Though it is possible to devise a special strategy only for these six cases, these cases are far less frequent than when the required voltage vector lies between two converter voltage vectors. Therefore, it is safe to continue with the vector sequence of Table 5.3 without facing a significant increase in switching losses.

In this section, we examined the Space Vector Pulse Width Modulation (SVPWM) in great detail. Though descriptions of SVPWM can be found in various forms in different texts of power electronics, quite often, the descriptions leave out the basic philosophy behind choosing and applying vectors. The previous sections had described the Clarke's transformation and how that can be used to generate a single rotating vector representation of a three-phase quantity. Furthermore, applying Clarke's transformation to the switching combinations of the three-leg converter of Fig. 5.4 produced the converter voltage vector hexagon. This section described how the voltage vector of the required three-phase output voltages can be superimposed on the converter voltage vector hexagon. Using several different examples, an algorithm for choosing the converter voltage vectors was described and eventually an algorithm was presented in the form of a flowchart. Following the choice of the converter voltage vectors, the time intervals for the vectors were calculated, and a vector sequence was presented for each vector combination. In the next section, we will present a simulation study for the three-leg converter of Fig. 5.4 controlled by SVPWM.

5.7 SVPWM Simulation

In the previous section, the process of SVPWM for the three-leg converter of Fig. 5.4 was described in detail. In this section, the theory and algorithm presented in the previous section, will be implemented in a simulation. The reader is strongly encouraged to try to simulate the three-leg converter controlled by SVPWM before accessing the simulation files by referring to the process described in the previous section. Moreover, the simulation package also contains a few variations on how the converter voltage vectors can be arranged in a sequence in a switching cycle. The reader is encouraged to think of other innovative ways to generate the vector sequence using the minimum amount of memory. Besides presenting the simulation results, this section will also utilize the vector diagrams of Figs. 5.12 and 5.14 to present a convenient manner to determine the capability of a three-leg converter to produce three-phase voltages.

The simulation of the three-leg converter of Fig. 5.4 with SVPWM can be found in the folder `three_leg_svpwm` within the folder `chapter5` at the link: https://github.com/opensourceelectrical/switching-strategies-for-power-electronics.

As stated in the previous section, there are a few modifications to how the precise vector sequences can be chosen, and these are made available as separate Python files which the reader is free to implement. The dc bus voltage V_{dc} has been chosen to be 24V, though in reality, a much larger voltage would be necessary to produce practical three-phase voltage supplies such as the 208V line-line RMS used in North America or the 400V line-line RMS used in Europe and many other parts of the world. Figures 5.16, 5.17 and 5.18 show the simulations results. Figure 5.16 shows the three-phase load currents which can be seen to balanced and sinusoidal with a large switching frequency ripple. Figure 5.17 shows the gate signals for the controllable devices S_1, S_3 and S_5 for a few switching cycles. Figure 5.18 shows the line-neutral voltages v_{an}, v_{bn}, v_{cn} that are applied across the three-phase load. These results are not very different from the results presented in the previous section that used sine-triangle bipolar PWM to control the same three-phase converter of Fig. 5.4.

In the simulation above, the dc bus voltage has been randomly chosen to be 24V, as the purpose of the simulation was merely to demonstrate the SVPWM strategy. However, an interesting by-product of the vector diagrams is that we now have a visual representation of

Fig. 5.16 Three-leg converter output currents

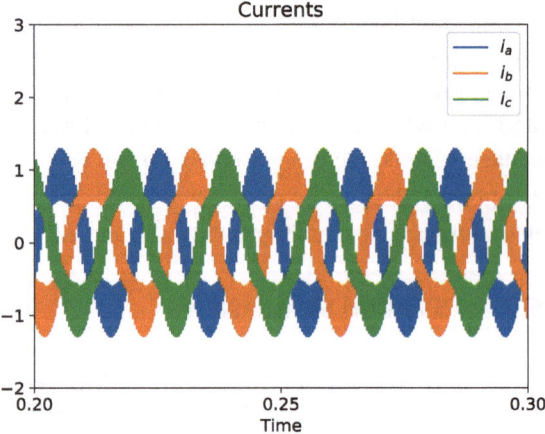

Fig. 5.17 Three-leg converter
PWM signals

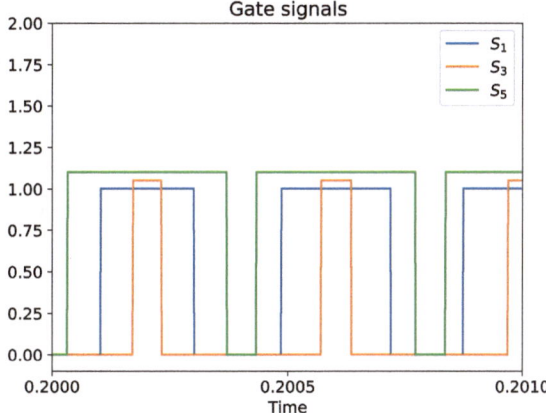

Fig. 5.18 Three-leg converter
line-neutral output voltages

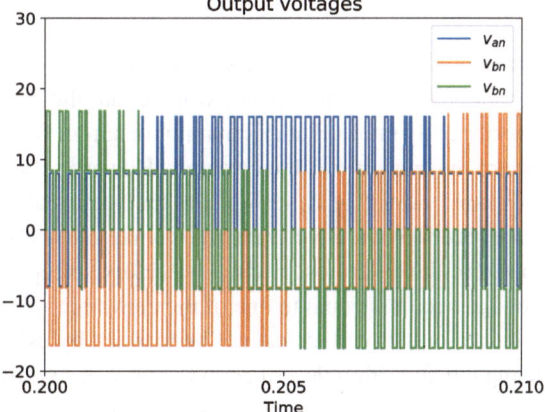

the capabilities of the converter. In the vector diagram of Fig. 5.14, we had superimposed the trajectory of the required output voltage vector that was obtained by performing Clarke's transformation on the sinusoidal expressions of the instantaneous output voltages. From Fig. 5.14, one can immediately conclude that the converter is capable of producing the required output voltage as the circular trajectory lies well within the magnitude of the converter voltage vectors. However, to fully understand the capabilities of the converter, let us complete the vector hexagon as shown in Fig. 5.19 with edges joining the converter voltage vectors. The Python code for generating this vector hexagon can be found in the file `converter_capacity.py` inside folder `vector_diagrams` within the folder `chapter5` at the link: https://github.com/opensourceelectrical/switching-strategies-for-power-electronics.

In the vector hexagon of Fig. 5.19, along with the circular trajectory of radius 0.3 which was plotted in Fig. 5.14, two more circular trajectories are plotted. The outer circular trajectory passes through the tips of all the converter voltage vectors, while the circular trajectory

Fig. 5.19 Vectors in the complex α-β plane for line-neutral voltages

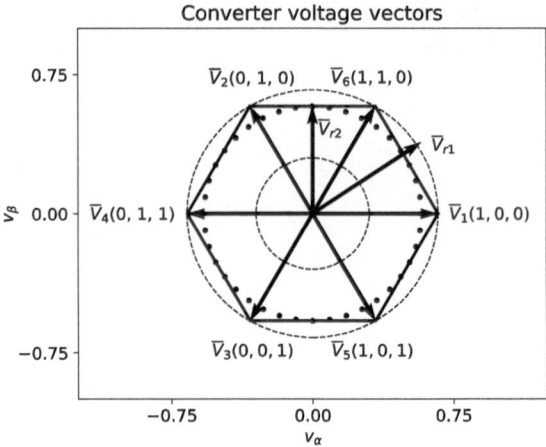

in the middle is such that every edge of the hexagon is tangential to it. When one is determining the maximum three-phase voltage that the three-leg converter can produce as outputs, one normally thinks that this maximum output voltage will be equal to the magnitude of any one of the converter voltage vectors. However, if one examines the outer circular trajectory, it is quite clear that it lies outside the vector hexagon. Therefore, if one attempts to produce required output voltages whose magnitude after Clarke's transformation is equal to the radius of the outermost circular trajectory, except for the instants when the required output voltage vector is equal to one of the converter voltage vectors, at all other time instants, the required output voltage cannot be realized by using the adjacent converter voltage vectors using the algorithm described in the previous sections.

As an example, let us consider some time instant t_1 such that $\omega t_1 = 30°$. If the required output voltages were chosen to be sinusoids:

$$v_{r1a} = 0.6667 \cos(\omega t)$$
$$v_{r1b} = 0.6667 \cos(\omega t - 120°) \tag{5.36}$$
$$v_{r1c} = 0.6667 \cos(\omega t - 240°)$$

The voltages obtained after performing Clarke's transformation would be:

$$v_{r1\alpha} = 0.6667 \cos(\omega t)$$
$$v_{r1\beta} = 0.6667 \sin(\omega t) \tag{5.37}$$

At $\omega t_1 = 30°$, the required output voltage vector $\overline{V}_{r1}$ would be as shown in Fig. 5.19.

Using the algorithm from the previous section and also by simple visual inspection, it is clear that converter voltage combination to be chosen is $(\overline{V}_1, \overline{V}_6)$. The next step would be to determine the time periods t_1 and t_2 for which the converter voltage vectors $\overline{V}_1$ and $\overline{V}_2$ need to be applied. We can use the expression (5.32):

$$0.6667 \times \cos(30°) \times T = 0.5773T = 0.6667 \times t_1 + 0.3333 \times t_2$$
$$0.6667 \times \sin(30°) \times T = 0.3333T = 0 \times t_1 + 0.5773 \times t_2 \tag{5.38}$$

Which can be solved to obtain:

$$t_1 = 0.5773\ T$$
$$t_2 = 0.5773\ T \tag{5.39}$$

From the above calculations, it quite clear that $t_1 + t_2 > T$; which is impossible. Therefore, the required output voltage that follows the outermost circular trajectory cannot be produced by the converter.

The limit of the converter can be determined by simple trigonometry, as we need the magnitude of the vector that is incident on the mid-point of any edge of the hexagon. A sample vector $\overline{V}_{r2}$ has been shown in Fig. 5.19. Since the converter voltage vectors are separated by 60°, the angle between the closest converter voltage vector and $\overline{V}_{r2}$ will be 30°. Therefore, the magnitude of $\overline{V}_{r2}$ can be expressed as:

$$|\overline{V}_{r2}| = 0.6667 \times \cos 30° = 0.5773 \tag{5.40}$$

Since we have assumed the dc bus voltage $V_{dc} = 1$ while generating the vector diagrams, the maximum RMS line-neutral output voltage that can be generated by the three-leg converter is:

$$V_{lnmax} = \frac{0.5773}{\sqrt{2}} V_{dc} = 0.4082 V_{dc} \tag{5.41}$$

And the maximum value of the line-line output voltages that can be generated is:

$$V_{llmax} = \frac{1.1547}{\sqrt{2}} V_{dc} = 0.8165 V_{dc} \tag{5.42}$$

It might be trivial to go through so much trouble merely to determine the maximum values of balanced three-phase output voltages that the three-leg converter can generate. However, in certain cases, we will need a converter to produce non-ideal balanced voltages which might be unbalanced or might contain harmonics. In such a case, plotting the trajectory of the required output voltages onto the vector diagrams provides a convenient procedure to determine the dc bus that would be necessary such that the converter can be suitable for the application. One can plot fairly complex trajectories by merely selecting sufficient number of samples in a cycle of the required output voltages, performing Clarke's transformation and plotting the coordinates $(v_{r\alpha}, v_{r\beta})$ as shown in the Python code in the file vector_algorithm_line_neutral.py inside folder vector_diagrams within the folder chapter5 at the link: https://github.com/opensourceelectrical/switching-strategies-for-power-electronics.

In this section, we examined the simulation results for the three-leg converter controlled by SVPWM. The results can be observed to be quite similar to those with sine-triangle PWM presented in a previous section. Along with the simulation results, we also used the converter voltage vector diagram to determine the voltage producing capacity of a three-leg converter. Though the capacity of any converter can always be determined analytically, a visual representation makes it convenient to interpret the results. Moreover, as any output voltage can be plotted as a trajectory on the converter voltage vector diagram, this visual tool can be quite easy to use when the converter is required to produced non-ideal or non-sinusoidal voltages. The next section will conclude this chapter.

5.8   Conclusions

This chapter presented in great detail, a procedure to implement SVPWM for a three-phase converter. The contents of this chapter are not very different from the previous chapters— the chapter presents a converter topology, examines the possible conduction modes and describes how a switching strategy can be formulated. However, given the multi-phase nature of the power converter topology, several other factors were introduced that need to be appreciated if the contents of this chapter are to become the basis of dealing with more complex topologies. In this conclusion, let us separate the contents of this chapter into similarities with the previous chapters, and into aspects where this chapter was very different from the previous chapters.

In terms of similarities, the first and foremost similarity is in the construction of the power converter. Though several different topologies are possible for three-phase converters, the topology used is one the simplest that is possible. The topology uses the single converter leg as a building block, which was also used in the previous chapters while dealing with the full-bridge and the half-bridge converter topologies. As stated before, the advantages of such a topology are the ease of construction in terms of compact modules. Fully functional modules are available from commercial suppliers of power devices that provide in a single module the three-phase converter comprised of three converter legs, in the same manner that the full-bridge and half-bridge topologies were available as modules. This makes it extremely convenient to build a three-phase converter in a hardware prototype.

Another similarity with respect to the previous chapters is the examination of all possible conduction modes of the three-phase converter. This examination helps any power electronics engineer understand all the possible conduction modes of the converter. In the case of the full-bridge and half-bridge topologies of the previous chapters, the output voltages of the converters could be immediately determined for each conduction mode. In the case of the three-phase converter, due to the multi-phase nature of the converter, the output voltages needed a certain amount of computation. However, listing the conduction modes of the converter is an extremely useful first step, as while formulating a switching strategy, the objective should be to utilize the conduction modes to result in a powerful and efficient converter that can produce the maximum possible output voltage while experiencing the lowest possible switching losses.

In terms of differences from the previous chapters, the first difference in this chapter was to represent the three-phase converter output voltage as a single vector. In the previous chapters on the full-bridge and half-bridge converters, since the output of the converters was a single voltage, the vector representation was much simpler. However, with a multi-phase output, a vector representation results in three rotating vectors which is not very convenient from the perspective of analysis. The Clarke's transformation was used to solve this hurdle, and transform the three-phase voltages into a another set of quantities. Due to the topology of the three-phase converter, the transformed values are reduced to two non-zero quantities that using vector diagrams as well as sinusoidal expressions were shown to have a special property, namely, that they formed a quadrature pair. This special property allows their representation in a separate complex plane which produces another rotating vector. This single rotating vector in the new plane of reference has become the single vector representation of the three-phase voltages.

The chapter described in great detail, how Clarke's transformation can be used to transform the outputs of the three-phase converter corresponding to every switching combination. Using Python code, the transformed voltages can be drawn as vectors, leading to a vector representation of the three-phase converter, whereby every switching combination of the converter corresponds to a unique vector. Though the process that leads to this vector diagram is drastically different from that used in the previous chapters, there are many common properties to this vector diagram. The vector diagram of the three-phase converter can be seen to have vectors of the same magnitude but equally separated from each other by an angle of 60°. Quite magically, it appears as if by using every possible switching combination, an entire cycle of three-phase output voltages can be realized.

In the previous chapter on full-bridge converters, though vector transitions were described and mapped to switching patterns, the vectors themselves were not directly used to generate the switching patterns. Due to the relative simplicity of the full-bridge converter, it was still possible to use a modified sine-triangle PWM technique to arrive at the required switching patterns. In this chapter, the SVPWM technique is demonstrated that allows us to choose converter voltage vectors directly in order to produce a certain desired output voltage. The SVPWM technique can be divided into several stages. The first stage being to produce a vector corresponding to the desired three-phase output voltages using Clarke's transformation, and superimpose this vector onto the converter voltage vector diagram. The second stage determines which of the converter voltage vectors need to be chosen in order to achieve the desired output voltage. The third stage determines the time periods for which each converter voltage vector that has been chosen will be applied in a switching cycle. The final stage determines the sequence of the converter voltage vectors such that the switching losses will be minimal.

The presentation of SVPWM has been broken up into several different stages, and also examined in great detail to provide the reader an in-depth understanding of the process. This process not only can be used in generating the switching signals for the three-phase converter, but can also be used as a tool to analyse the converter and understand the capacity of the converter. The reader is strongly encouraged to try solving equations on their own and also to run the simulations and code presented in this chapter. The analysis and code used in this chapter will be used in the next chapter on multi-level converters.

References

1. S.J. Chapman, *Electric Machinery Fundamentals* (McGraw-Hill, 2004)
2. C.K. Alexander, M.N. Sadiku, M. Sadiku, *Fundamentals of Electric Circuits* (McGraw-Hill Higher Education Boston, 2007)
3. J.P. Agrawal, *Power Electronic Systems. Theory and Design* (Prentice Hall, Upper Saddle River, 2001)
4. K. Taniguchi, Y. Ogino, H. Irie, PWM technique for power MOSFET inverter. IEEE Trans. Power Electron. **3**(3), 328–334 (1988)
5. "Clarke & Park transforms on the TMS320C2xx," Texas Instruments, Technical Report (1997). [Online]. Available: https://www.ti.com/lit/an/bpra048/bpra048.pdf
6. W.C. Duesterhoeft, M.W. Schulz, E. Clarke, Determination of instantaneous currents and voltages by means of alpha, beta, and zero components. Trans. Am. Inst. Electr. Eng. **70**(2), 1248–1255 (1951)
7. C.J. O'Rourke, M.M. Qasim, M.R. Overlin, J.L. Kirtley, A geometric interpretation of reference frames and transformations: dq0, Clarke, and Park. IEEE Trans. Energy Convers. **34**(4), 2070–2083 (2019)
8. L. Umanand, *Power Electronics: Essentials and Applications* (Wiley India Pvt. Limited, 2009)
9. A.K. Chakraborty, B. Bhattachaya, Determination of α, β and γ-components of a switching state without Clarke transformation, in *2016 2nd International Conference on Control, Instrumentation, Energy & Communication (CIEC)* (IEEE, 2016), pp. 260–263
10. P. Randewijk, *An overview of space vector PWM*, Technical Report (2004)

Multi-level Converters 6

6.1 Introduction

The previous chapter introduced three-phase systems which are commonly used in the power system and also in industry. For three-phase systems, the chapter presented a basic three-leg converter topology with each leg serving as the output terminal for a phase. Most importantly, the chapter described how three-phase quantities can be expressed as vectors after transforming them using Clarke's transformation, and using this procedure, how the output voltages of the three-leg converter can be represented using a vector diagram. The chapter described in detail the approach of SVPWM as a switching strategy. Using SVPWM results in optimal usage of a converter with the maximum possible output voltages and the least switching frequency due to the systematic approach in choosing the switching combinations that will best result in a required output voltage. In this chapter, we will use these concepts in the domain of high power converters.

In exactly the same manner that three-phase systems were introduced to increase the power handling capability of ac systems, power engineers began examining ways to increase the power rating of power converters. The power rating of a converter is limited by the rating of the power devices that form the converter. In terms of the rating of the power devices, there are many factors that need to be taken into account. The forward blocking voltage and the forward current of a power device are the maximum voltage that the device can block in the absence of a gating signal and the maximum current that can pass through the device in the presence of a gating signal. The turn ON and turn OFF times are the minimum time interval needed for the device to start conducting after the application of a positive gate signal and the minimum time interval needed for the device to stop conducting after the application of a negative gate signal. Power devices are available in several ratings. There are low power MOSFETs which are best suited for low power and low voltage dc-dc applications that can pass forward currents of a few tens of amperes and block a voltage of around a hundred volts. There are high power IGCTs that can pass hundreds of amperes of current and block several hundreds or a few thousands of volts of forward voltage. Low power devices usually have very low turn ON and turn OFF times in the range of a few nanoseconds while high power devices can take a few microseconds to turn ON and turn OFF.

High power applications can either be at medium voltage such as between around 690V to a few kilovolts but with very high current ratings in either hundreds or even thousands of amperes, or can be at very high voltage such as 11–220 kV with currents in the range of a few hundred amperes. To be able to build a converter for such ratings, quite often one needs to connect several high power devices in series and parallel such that the cumulative blocking voltage and current handling capability increases with the number of devices being connected. Furthermore, devices that handle such high voltages and currents cannot be turned ON and OFF very rapidly, as the switching losses that result can cause unacceptable temperature rises and damage the devices. For these reasons, designing high power converters is extremely challenging. One can find books and research papers dedicated to the design and control of high power rating converters. Since this chapter is the concluding chapter in this book that was intended to give the reader a flavour of power electronics relevant to industry, this chapter will introduce one solution that is commonly used for designing high power converters, namely the multi-level converter.

This chapter will describe how the levels of any converter can be defined. With respect to this definition, the converters examined in the previous chapters were 2-level converters. This chapter will describe how one can modify the 2-level converter leg used in the previous chapter to a multi-level converter leg. The chapter will use an iterative and questioning approach on how the popular Neutral Point Clamped (NPC) topology came into being, and how any k-level converter can be designed using the NPC strategy. The chapter will describe the operation of a NPC multi-level converter by defining the forbidden conduction modes of operation, and subsequently defining the allowable conduction modes. Using the similar method of permutations, the total allowable conduction modes can be determined. The chapter will present specific discussions related to the 3-phase 3-level and 3-phase 4-level converters which are quite popular for medium voltage applications.

The chapter will use Python code to generate the allowable modes of conduction, and for each conduction mode calculate the output voltage produced. These output voltages will be transformed using Clarke's transformation to produce vectors corresponding to each switching combination. These vectors can be represented in a vector diagram similar to the previous chapter, except that the vector diagrams of multi-level converters will be comprised of a much larger number of vectors. The SVPWM strategy for the multi-level converter will be described in-depth using the example of a 3-phase 3-level converter. The algorithm for choosing converter voltage vectors will be described, followed by calculation of the time intervals for which the voltage vectors will need to be applied. The chapter will also describe how the redundancy in the vectors can be utilized to balance the voltages of the dc bus capacitors.

This chapter will focus specifically on the 3-phase 3-level converter with a few references to the 3-phase 4-level converter. This has been done to be able to describe in detail how

SVPWM can be implemented for a multi-level converter. This is in contrast to most other books that will deal with a vast range of multi-level converter topologies making it very difficult for a newcomer to power electronics to fully understand the working of these converters. Instead this chapter uses discussion and intuitive thinking to describe how a multi-level converter can be constructed and operated. The reader is strongly encouraged to examine the simulations along with the theory in this chapter, as due to the complexity of these converters, merely reading the text in this chapter will not be sufficient to understand their operation.

6.2   Overview of Multi-level Converters

In this section, we will introduce the concept of multi-level converters without actual details on the converters themselves. The description of the topology, the conduction modes and switching strategies will be deferred until the later sections. This section will address the question about why multi-level converters are needed and when would the converters presented in the previous chapters not be sufficient. To answer these questions, one needs only a basic understanding of how a power converter functions, and some background on industrial applications that make these multi-level converters necessary.

In the past few chapters, the single converter leg comprising of two controllable devices and their anti-parallel diodes is used as a building block for synthesizing power converters. Power electronic converters for any application can be realized in many different ways, but as shown in the Chap. 3, a single converter leg can fulfil most converter requirements while also providing the convenience of using a compact commercially available power module that contains not only the power devices but also auxiliary circuits such as gate drivers and protection circuits. When we speak of a converter leg, once again, there are many variations that are commercially available. Until now we only considered the converter leg with two controllable devices and two anti-parallel diodes. This topology is repeated in Fig. 6.1. Let us suppose that the extreme terminals of the converter leg are connected across a dc voltage source. Though such a circuit alone is of no use, it will serve to describe the concept of levels in the output voltage.

If we define the positive and negative terminals of the dc voltage source as P and N respectively as shown in Fig. 6.1, we can define a voltage v_{oN} between the mid-point of the converter leg and the negative terminal of the dc voltage source. For the converter leg of Fig. 6.1, it is very clear that the voltage v_{oN} will have two levels, namely V_{dc} and 0. The voltage v_{oN} will have a value of V_{dc} when the upper devices are conducting and a value of 0 when the lower devices are conducting. If this converter leg is used for a half-bridge converter, as described in Chap. 3, the output of the converter turned out to be $\frac{V_{dc}}{2}$ or $-\frac{V_{dc}}{2}$ due to the neutral of the ac system connected to the mid-point of the dc bus capacitors. If two of these converter legs are used in a full-bridge converter, as described in Chap. 4, the output of the converter turned out to be V_{dc}, 0 or $-V_{dc}$ as the two converter legs can be

Fig. 6.1 Converter leg with
two devices

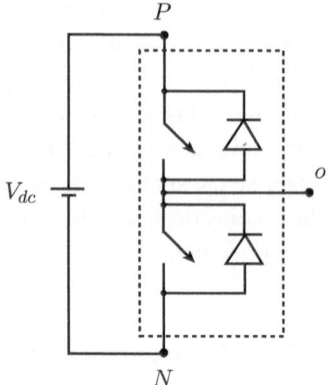

independently regulated. In the case of the three-phase converter, as described in Chap. 5, the output voltages can have many different levels depending on the connection of the ac system being supplied by the converter.

At first glance it appears that using a single converter leg shown in Fig. 6.1, it is possible to achieve several levels in the final output voltage. However, it is important to note that the output voltage produced is a combination of the interaction between multiple converter legs and also the rest of the ac system. If we need to define levels for any converter, we must be able to do so without taking these additional factors into account, as they can always change. Therefore, we are asking the question, for a given basic building block, how many voltage levels can a converter produce. If one looks at Fig. 6.1, by defining the voltage v_{oN}, we are in a position to answer this question without taking into account any other converter legs or how the converter is connected to the ac system. With the converter leg in this sandbox, we can conclude that a converter comprising of these converter legs can produce two voltage levels. Hence, this converter is called a two-level converter.

In Chap. 2, while describing the challenge in power electronics, we examined the case where a single switch was regulated to produce a sine wave. By using PWM, the switch status can be regulated such that the pattern formed contains a fundamental frequency corresponding to the required sine wave and higher order harmonics equal to the frequency of carrier waveform. Due to the fact that the switch can attain only two states (ON and OFF), the only efficient manner of producing a sine waveform was to vary the width of the pulses of the switch states. The switching frequency component that is also produced can be removed by a low-pass filter. In the simulations of the previous chapters, we examined voltage patterns produced by converters, and also the output currents that had besides a grid frequency sine waveform, high frequency ripples. Since the high frequency components are a by-product of PWM, we would like to design a low-pass filter that usually comprises of inductors and capacitors that attenuates the high frequency ripple to a negligible amount.

While designing converters for practical applications, the objective would be to design a system that satisfies the requirements while also minimizing cost. The cost of the system is determined by the cost of the converter components, the low-pass filter and auxiliary components related to protection, sensing, communication etc. To decrease the cost, one would choose power devices that can withstand the currents and voltages expected with adequate safety margins, and achieve the switching frequency chosen for the converter. Furthermore, one would also like to decrease the cost of the low-pass filter components while also ensuring that they achieve the objective of minimizing the switching frequency ripple. For low power converters, this is usually not a very challenging optimizing problem. However, as the power rating of the system increases, this challenge can become fairly daunting.

As the power rating increases, the switching losses and conduction losses in the converter power devices will increase. Beyond a threshhold, it might be necessary to use multiple power devices connected in series or parallel instead of just a single power device. It might also become necessary to select specially constructed power devices or another family of devices based on another technology or material. With respect to the low pass filter, high power ratings translate into higher currents and voltages, which implies inductors with larger current rating and capacitors with higher voltage rating. This will lead to higher costs of the filter, and in the case of high current inductors, will also result in an increase in size as the inductors usually have metal (laminated iron quite often) cores that are the main reason for the bulk. There are several approaches to alleviate this problem.

One could decrease the switching frequency of the converter, which will lead to lower switching losses in the devices. However, a decrease in switching frequency will bring it closer to the grid frequency. This in turn will result in an increase in the size of the filter components. In many applications, it is usually not acceptable to increase the size of the low-pass filter components, and therefore one needs to find a way to maintain the same switching frequency while still reducing the number of transitions of the power devices in the converter. One way to achieve this is by increasing the number of levels in the output voltage produced. If we try to achieve a sine waveform with a switch that can achieve two states, the only variable that we can control is the width of the pulses of the ON and OFF periods. However, if we bring in more than two states, besides the width of the pulses, the levels can be so arranged that the switched waveform is closer to the required sine waveform.

The exact details of how this will result in decreased switching transitions while still maintaining the effective switching frequency is something that is best described using detailed analysis, and the remaining sections will cover that. Before closing this section and beginning with the topology of a multi-level converter, let us describe briefly the different types of multi-level converters. As with all power electronics converters, multi-level converters can be achieved in many different ways. One of the most intuitive methods would be to merely connect full-bridge converters in series with each converter having the same dc bus voltage, such that the final output voltages is the sum of all the output voltages of all the converters. These type of converters are called cascaded converters [1]. Since it is already known that a full-bridge converter can produce three levels of output voltage, cascading k full-bridge converters will result in $2k + 1$ levels in the final output voltage. As an example, for three full-bridge converters connected in series, the final output voltage will have the levels $3V_{dc}$, $2V_{dc}$, V_{dc}, 0, $-V_{dc}$, $-2V_{dc}$ and $-3V_{dc}$.

Another method to construct the multi-level converter is called the Neutral Point Clamped (NPC) multi-level converter [2]. This structure is an extension of the two-level converter leg above except that this requires the dc bus to have intermediate tappings which is usually achieved by connecting series electrolytic capacitors across the dc voltage source and extracting a terminal at each connection. Additional power devices will allow the output terminal of the leg to be connected to the upper terminal P of the dc bus, the intermediate terminals of the dc bus or the lower terminal N of the dc bus. As a result, for k capacitors connected in series to form the dc bus, the multiple levels that can be achieved will be kV_{dc}, $(k-1)V_{dc}$, ..., V_{dc} and 0. To achieve this, one requires a converter leg with a greater number of controllable devices and their associated anti-parallel diodes besides also auxiliary devices that connect the leg and the dc bus capacitors [2]. Detailed descriptions will follow in the coming sections.

Another method to construct a multi-level converter is called the Flying Capacitor (FC) multi-level converter [3]. In this case, one uses a number of capacitors in addition to the dc bus capacitors to form intermediate voltage levels. These converters are fairly complex and need a good deal of voltage balancing between these capacitors for the topology to be effective [4, 5]. In this chapter, only the NPC multi-level converters will be described as though a certain degree of voltage balancing is needed, the switching strategy can be described without delving too much into voltage balancing control loops. In the next section, we will examine the topology of the 3-level and 4-level NPC converters.

6.3   Multi-level Converter Topology

The previous section provided an overview of multi-level converters, the reason for choosing a multi-level topology and a few different types of multi-level converters. In this section, we will examine in detail the topologies of the 3-level and 4-level NPC converters. One can find the usage of 5-level NPC converters also in high-power motor drive applications and in some cases, also in High Voltage DC (HVDC) applications. Though there is no theoretical limit to the number of levels a multi-level NPC converter can achieve, levels greater than five are usually rare due to the greater number of power devices as well as the complexity of the control. The objective of this section is to present a step-by-step approach to realize a multi-level NPC converter.

Figure 6.2 shows how one can gradually realize a 3-level converter [2]. To begin with, once we know the number of levels, this determines the number of electrolytic capacitors that need to be connected in series across the dc voltage source. For a k level converter, one needs $k-1$ series-connected electrolytic capacitors with $k-2$ number of intermediate terminals. This is clear from (a) of Fig. 6.2, where two electrolytic capacitors are used for a 3-level converter and the mid-point M of these capacitors becomes an important terminal along with

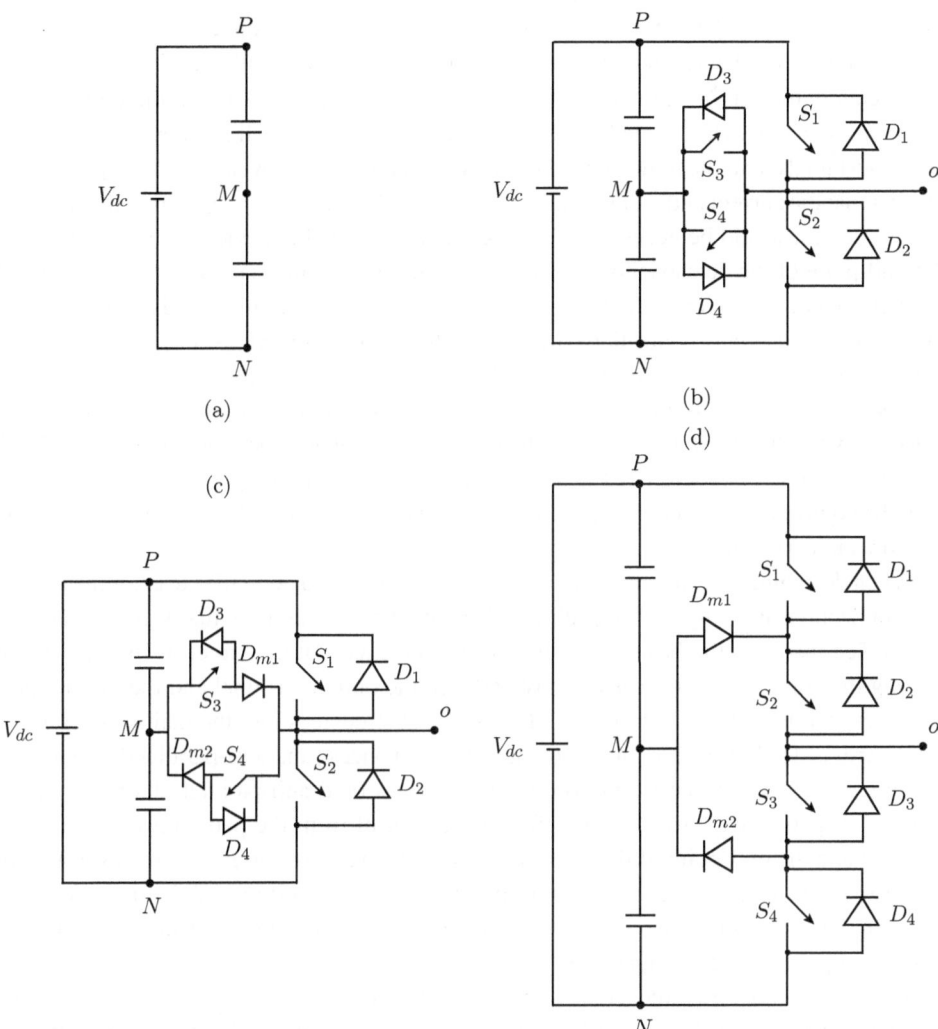

Fig. 6.2 Three level converter leg evolution

P and N. The next step is to connect across the dc bus a leg of controllable devices along with their associated anti-parallel diodes. The output terminal of the leg will be extracted from the mid-point of the devices which implies that the number of controllable devices (and anti-parallel diodes) will always be an even number. For the output of the converter leg to be connected to the positive terminal P and produce an output voltage of V_{dc}, all the upper devices must conduct, while for the output of leg to be connected to the negative terminal N and produce an output voltage of 0, all the lower devices must conduct.

We have not yet determined the number of devices in the converter leg. This can be answered once we determine how to achieve an output of $\frac{V_{dc}}{2}$. Quite obviously, in order to achieve an output of $\frac{V_{dc}}{2}$, the output terminal o needs to be connected to the mid-point M of the dc capacitors. One immediate thought is to connect a controllable device between the output terminal o and the mid-point M of the dc capacitors. However, in order to account for the fact that the current at the output terminal could either be leaving the terminal or entering the terminal, it would be necessary to connect two controllable devices S_3 and S_4 between the mid-point of the dc capacitors and the output terminal. This is shown in (b) of Fig. 6.2. Though we need only controllable devices between the output terminal and the mid-point of the dc capacitors, many controllable devices such as IGBTs and MOSFETs have an in-built body diode which is in anti-parallel across the controllable device as shown in (b) of Fig. 6.2. It is essential that we disable these body diodes D_3 and D_4 or else an uncontrollable path exists between the output terminal and the mid-point of the dc capacitors. In order to disable the body diodes, we would need to connect another pair of diodes D_{m1} and D_{m2} in series with the controllable devices in such a manner that only the controllable device can conduct as shown in (c) of Fig. 6.2.

When designing a power electronic converter, it is always a good idea to examine alternative connections that might also result in a functional converter. Eventually, the final design is a choice made on the basis of efficiency, convenience, safety and cost. Instead of connecting the controllable devices directly between the output terminal and the mid-point of the dc capacitors, what if we inserted the controllable devices into the main converter leg. Therefore, instead of having only two controllable devices (and anti-parallel diodes) in the converter leg, we now have four controllable devices (and anti-parallel diodes) as shown in (d) of Fig. 6.2. We have now transferred the control from the path between the output terminal and the mid-point of the dc capacitors to the main converter leg. However, we still need to establish a path between the mid-point of the dc capacitors and the main converter leg such that, when either device S_2 or S_3 are conducting, the output terminal is connected to the mid-point of the dc capacitors.

This path cannot be a direct connection or it can lead to a short-circuit of the dc capacitors when the devices S_1 or S_4 are conducting. This path must facilitate the flow of current when devices S_2 or S_3 are conducting, but should not permit devices S_1 or S_4 to short-circuit the dc capacitors. Such conditional flow of current can be achieved by connecting diodes D_{m1} and D_{m2} between the mid-point of the dc capacitors and the main converter leg as shown in (d) of Fig. 6.2. The device S_2 can conduct a current only in the downward direction which would occur when the output current is leaving the output terminal, in which case diode D_{m1} will conduct and supply the current either to the upper capacitor or from the lower capacitor. The device S_3 can conduct a current only in the downward direction which would occur when the output current is entering the output terminal, in which case diode D_{m2} will conduct and supply current either from the upper capacitor or to the lower capacitor. The reader should verify that when device S_1 conducts, D_{m1} will be reverse biased, and when device S_4 conducts, D_{m2} will be reverse biased.

The question arises—which is the better topology - (c) or (d) of Fig. 6.2? From the perspective of simplicity (c) is a simpler option as the connection of the additional controllable devices and diodes is intuitive and fairly obvious. However, from the viewpoint of convenience, it would be more convenient to have all controllable devices and their anti-parallel diodes connected in series in one leg. One would still need to ensure that all the controllable devices are not turned ON simultaneously which would short-circuit the dc bus. In the case of (c), not only would do we need to ensure that the upper device and lower device are not turned ON simultaneously, it would also be essential that when the devices connecting the output terminal to the mid-point of the dc capacitors are turned ON, the controllable devices in the main leg are turned OFF, or else one of the dc capacitors could be short-circuited. Therefore, from the perspective of control complexity, (c) of Fig. 6.2 will need the definition of another forbidden conduction mode.

Since multi-level converters are usually for high power applications, the output voltages and currents of the converter will be much higher. Therefore, in order to achieve the higher output voltages, it might be necessary to choose a high voltage dc bus. When lower devices are conducting, the upper devices will need to withstand a voltage of $\frac{V_{dc}}{2}$. In the case of (c) of Fig. 6.2, a single device S_1 will need to withstand $\frac{V_{dc}}{2}$, whereas in the case of (d) of Fig. 6.2, two devices S_1 and S_2 will need to withstand $\frac{V_{dc}}{2}$ which implies that each device will need to withstand approximately $\frac{V_{dc}}{4}$. Therefore, from the perspective of device forward blocking capability, the topology of (d) in Fig. 6.2 would need less expensive devices as compared to the topology of (c) in Fig. 6.2. In such a manner, one can draw up a long list of pros and cons between the two topologies. For commercial industrial applications, the process can be much more complicated as it also becomes necessary to mass manufacture Printed Circuit Boards (PCBs), source components and fabricate the final product.

The above description was merely to give a newcomer to power electronics a flavour of converter design, as rather than just presenting a converter topology and describing its operation, from the perspective of learning a new topic, it is much more interesting to examine how one could experiment with topologies and ponder about how and why a topology would function or not function. Moreover, it is also important to examine how any change to a topology in terms of addition or removal of a branch or component can affect the performance of the converter. It is always possible that adding a branch or component may result in the possibility of a short-circuit or might need a conditional check to ensure that two devices do not conduct simultaneously. Sometimes to prevent a short-circuit or the flow of current in a direction, it might be necessary to add a diode or a controllable device. Though a lot of these design considerations might seem random and heuristic, it also adds to the fascination that power electronics brings with it, as there is always the possibility of modifying a topology to suit a purpose.

The 3-level converter leg of Fig. 6.2 is surprisingly extendible and can be used to synthesize higher levels of multi-level converters. As an example, the 4-level converter leg can be synthesized as shown in Fig. 6.3 [2]. One can follow the same logic of constructing a 3-level converter leg as shown in Fig. 6.2. To begin with, for a 4-level converter ($k = 4$), one

Fig. 6.3 Four level converter leg topology

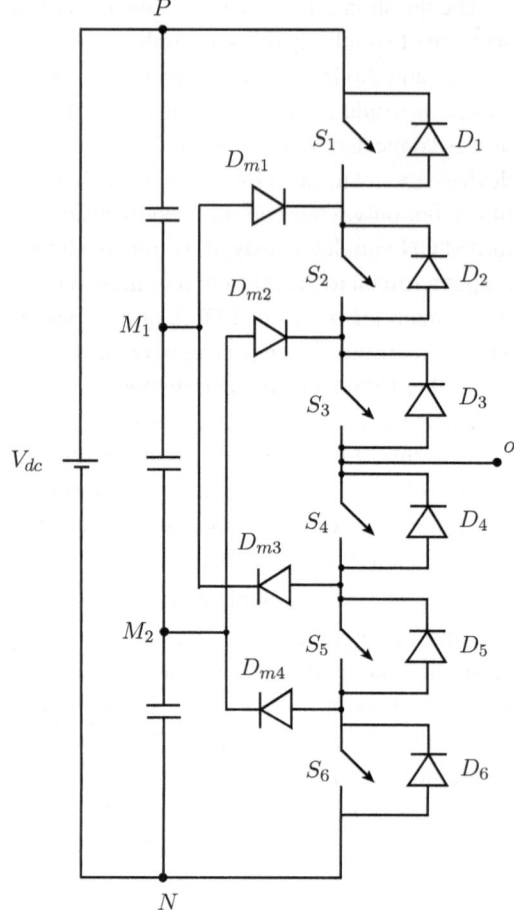

would need $k - 1 = 3$ electrolytic capacitors resulting in $k - 2 = 2$ interconnection points M_1 and M_2. To determine the number of devices in the converter leg, one needs to follow the following logic. To connect the output terminal to one of the interconnection points of the capacitor, one needs a pair of controllable devices. This is essential as the output current can be leaving or entering a terminal, and therefore, one needs a controllable device that can conduct in each direction. With two interconnection points, one has $2 \times (k - 2) = 4$ intermediate controllable devices (and their anti-parallel diodes). Finally, there are two extreme controllable devices that connect to the positive terminal P and the negative terminal N of the dc bus. This makes the total number of controllable devices (and their associated anti-parallel diodes) to be $2 \times (k - 1) = 6$.

However, one must take into consideration a few factors while choosing these intermediate controllable devices. The pair of intermediate controllable devices are connected such that one of them is above the mid-point of the converter leg, while the other is below the converter leg. In the case of the 3-level converter, both controllable devices were connected to the output terminal due to the fact that there were four controllable devices altogether in the leg. But with six controllable devices, quite obviously, two of the controllable devices will not be directly connected to the mid-point output terminal. This can be seen in Fig. 6.3 where S_2 and S_5 have no direct connection to either the mid-point or to the extreme terminals P or N. Therefore, if one wishes to achieve a connection from the output terminal to an intersection terminal, this may have to be achieved through the conduction of multiple devices. As an example, to connect the output terminal o to intersection terminal M_1, either devices S_2 and S_3 will need to conduct or device S_4 will need to conduct depending on the direction of the output current. The diodes D_{m1}, D_{m2}, D_{m3} and D_{m4} merely establish conditional paths and avoid the short-circuit of the dc bus capacitors when S_1 or S_6 are conducting.

The reader is encouraged to draw the topology of the 4-level converter on a piece of paper and figure out the operation. If we wish the output v_{oN} to be V_{dc}, all the upper devices S_1, S_2 and S_3 (or D_1, D_2 and D_3) must conduct. If we wish the output v_{oN} to be $\frac{2}{3}V_{dc}$, we would need devices S_2 and S_3 to conduct along with diode D_{m1} or device S_4 to conduct along with D_{m3} depending on the direction of the output current. If we would like the output v_{oN} to be $\frac{1}{3}V_{dc}$, we would need devices S_3 to conduct along with diode D_{m2} or devices S_4 and S_5 to conduct along with D_{m4} depending on the direction of the output current. If we would like the output v_{oN} to be 0, all the lower devices S_4, S_5 and S_6 (or D_4, D_5 and D_6) must conduct. With four possibilities to the output voltage v_{oN}, namely V_{dc}, $\frac{2}{3}V_{dc}$, $\frac{1}{3}V_{dc}$ and 0, quite clearly, the converter leg of Fig. 6.3 can be the basic building block for a multi-phase 4-level converter. In this same manner, one can also use the same logic to synthesize a 5-level converter, which will be left as an exercise to the reader.

This section described to a reader how it is possible to synthesize a multi-level converter leg using basic logic and reasoning. Most newcomers find it frustrating how power electronics sometimes seems random and heuristic with a large number of topologies available for a single application. However, to understand a topology, or to modify a topology, one only needs to understand the basic principles of non-linear conduction paths, and how these can be formed using controllable devices and/or diodes. Though one can think of innovative methods to synthesize a converter, using a systematic approach has its benefits as one can use building blocks and commercially available modules for the purpose. Though the basic operation of 3-level and multi-level converters has been described here along with the topology, the next section will describe in detail the conduction modes which in turn will allow us to represent the converter voltages as vectors.

6.4   Operation of a Multi-level Converter

In the previous section, we had examined how a multi-level converter can be constructed. The topology of the converter was described along with the basic operating philosophy, as when one understands the requirement, the approach towards fulfilling that requirement becomes clearer. As already stated, out of the many different ways in which multi-level converters can be constructed, the Neutral Point Clamped (NPC) type of converters will be discussed. Multi-level converters have been the topic of intense research for the past few decades, and therefore attempting to cover them in detail will require a book dedicated to multi-level converters. However, the past section presented NPC type 3-level and 4-level converter leg topologies in a manner such that the reader can imagine an extension towards higher level converters as well. In this section, the operation of the multi-level converter legs will be described in a similar abstract manner which can be extended towards other higher level converters.

In order to fully understand the operation of a converter, it is necessary to list out the conduction modes as well as the forbidden conduction modes. Such an approach ensures that any switching strategy developed fully utilizes the capacity of the converter while ensuring safety. Before listing the conduction modes of the 3-level and 4-level converter leg topologies of Figs. 6.2 and 6.3 respectively, let us begin by examining the forbidden conduction modes. One obvious conduction mode that should not be allowed is when all the controllable devices in a single leg are conducting simultaneously as this would short-circuit the dc voltage source. In the case of a 2-level converter, one can conveniently state that the conduction of the upper and lower devices should be complementary to avoid such a short-circuit. However, for multi-level converter legs with several devices, such a simple rule cannot be devised as there could always be some conduction state that needs a few upper devices and a few lower devices to conduct as has already been described in the previous section. Therefore, one needs to ask the question - what condition should cause a conduction mode to be forbidden?

Quite simply, any conduction mode that results in the dc voltage source or any one or more of the electrolytic capacitors to be short-circuited should be a forbidden state. The question then arises, when would a capacitor be short-circuited? If one examines the 3-level converter leg of Fig. 6.2, it is possible to think of a few conduction modes where the capacitors could become short-circuited. For the upper capacitor to be short-circuited, devices S_1, S_2 and S_3 would need to conduct, and in that case it is easy to see that a short-circuit path is formed with the diode D_{m2}. For the lower capacitor to be short-circuited, the devices S_2, S_3 and S_4 would need to conduct, and a short-circuit path is formed with diode D_{m1}. The reader is strongly encouraged to take a paper and pencil and draw the loops that result from these conduction modes to verify that capacitors will be short-circuited.

In a similar manner, by examining the 4-level converter of Fig. 6.3, one can find similar paths that will short-circuit each of the three capacitors. For the upper-most capacitor to be short-circuited, the devices S_1, S_2, S_3 and S_4 would need to conduct and a path will be formed along with diode D_{m3}. For the middle capacitor to be short-circuited, the devices S_2, S_3, S_4 and S_5 would need to conduct and a path will be formed along with diode D_{m1} and D_{m4}. For the lower-most capacitor to be short-circuited, the devices S_3, S_4, S_5 and S_6 would need to conduct and a path will be formed along with diode D_{m2}. Since it also possible for multiple capacitors to be short-circuited, one can observe that the series combination of the upper and middle capacitor will be short-circuited when devices S_1, S_2, S_3, S_4 and S_5 are conducting as a path will be formed with diode D_{m4}. Similarly, the middle and lower capacitors will be short-circuited when devices S_2, S_3, S_4, S_5 and S_6 are conducting as a path will be formed with diode D_{m1}. However, the short-circuit of multiple capacitors is merely a subset of the short-circuit of a single capacitor, and if one defined forbidden conduction modes that result in the short-circuit of a single capacitor, that will also prevent the short-circuit of multiple capacitors.

After defining the above forbidden modes for the 3-level converter and the 4-level converter, one can make a few observations. For a 3-level converter, the conduction of more than two controllable devices will lead to a short-circuit of either one of the capacitors or the entire dc voltage source. For a 4-level converter, the conduction of more than three controllable devices will lead to a short-circuit of either one of the capacitors, two capacitors as a series combination or the entire dc voltage source. The next question, what switching combinations will be of use and what could be termed as useless? This question needs to be asked in the case of multi-level converter legs consisting of several devices as in Figs. 6.2 and 6.3, as it is clear that in many cases more than a certain number of devices must conduct in order that the output terminal be connected to either the positive or negative terminals of the dc bus or to an intersection terminal of the electrolytic capacitors.

For a 3-level converter, it is clear that if only one controllable device conducts, it will be of no use. As an example, if only S_1 or S_4 conduct, this will not *intentionally* connect the output terminal to either positive P or negative terminal N, as the intermediate devices are not conducting. The stress is made on the word intentionally, as if only one device is conducting, the current will find its way through the anti-parallel diodes. If the current were leaving the output terminal, and if S_1 or S_4 were turned ON, the current would flow through diodes D_3 and D_4. The current cannot flow through S_1 as for that to happen, device S_2 must also be turned ON and conducting. The current cannot also flow through S_3 and S_4 as in that case the current would be flowing against the direction of conduction. If the current were entering the output terminal, it would flow through diodes D_1 and D_2 for the same reason—cannot flow through S_4 unless S_3 also conducts, and cannot flow through S_1 and S_2 as it is in the opposite direction [6].

If only S_2 or S_3 are turned ON, it does not guarantee that the output terminal will be connected to the mid-point M as a device can only conduct for a particular direction of current. To ensure connection of the output terminal o to the mid-point M, both devices S_2 and S_3 must be turned ON, in which case depending on the direction of current, one of the devices will conduct along with one of the diodes D_{m1} or D_{m2}. If the current were leaving the output terminal, if only S_2 were turned ON, the output would be connected to the mid-point through the diode D_{m1}, though this is merely good fortune. If S_2 were not turned ON, and S_3 was turned ON, the current could not flow through S_3 due to it being in the opposite direction, but will instead flow through D_3 and D_4. If the current were entering the output terminal, if only S_3 were turned ON, the output would be connected to the mid-point through the diode D_{m2}, though as before, this is merely good fortune. If S_3 were not turned

Table 6.1 3-level converter leg output voltage corresponding to switching combinations

S_1	S_2	S_3	S_4	v_{oN}
OFF	OFF	ON	ON	0
OFF	ON	ON	OFF	$\frac{V_{dc}}{2}$
ON	ON	OFF	OFF	V_{dc}

Table 6.2 4-level converter leg output voltage corresponding to switching combinations

S_1	S_2	S_3	S_4	S_5	S_6	v_{oN}
OFF	OFF	OFF	ON	ON	ON	0
OFF	OFF	ON	ON	ON	OFF	$\frac{1}{3}V_{dc}$
OFF	ON	ON	ON	OFF	OFF	$\frac{2}{3}V_{dc}$
ON	ON	ON	OFF	OFF	OFF	V_{dc}

ON, and S_2 was turned ON, the current could not flow through S_2 due to it being in the opposite direction, but will instead flow through D_1 and D_2. Therefore, in the case of a 3-level converter leg, two devices must be turned ON simultaneously for any intentional output, or else depending on the direction of current, the output terminal will be connected to either positive P terminal or negative N terminal of the dc bus [7].

The reader is encouraged to perform this exercise for the 4-level converter, and will arrive at a similar but different conclusion, namely that for an *intentional* output to be produced, three devices need to be turned ON, or else the output terminal will be connected to the positive P terminal through diodes D_1, D_2 and D_3 or to the negative N terminal through diodes D_4, D_5 and D_6 depending on the direction of the current. Since we have determined that for a 3-level converter leg, two controllable devices need to be turned ON simultaneously, and for a 4-level converter leg, three devices need to turned ON simultaneously, one can list the conduction modes for them as shown in Tables 6.1 and 6.2. For a 3-level converter leg, this results in three conduction modes, and for a 4-level converter leg, it results in four conduction modes.

Similar to the 2-level converter, a multi-level converter leg can be used for any application. It can be used as a half-bridge for a low power dc-ac converter, or two multi-level converter legs can be used for either dc-ac or dc-dc applications, or three multi-level converter legs can be used for three-phase applications. Though vast literature can be found that uses multi-level converter legs in various applications, the most common applications are in three-phase applications either in industrial applications that need high power, or in power system applications at medium voltage or high voltage that are also high power applications. Figure 6.4 shows a three-phase converter using 3-level converter legs. The reader is encouraged to draw the topology of a three-phase converter using 4-level converter legs as

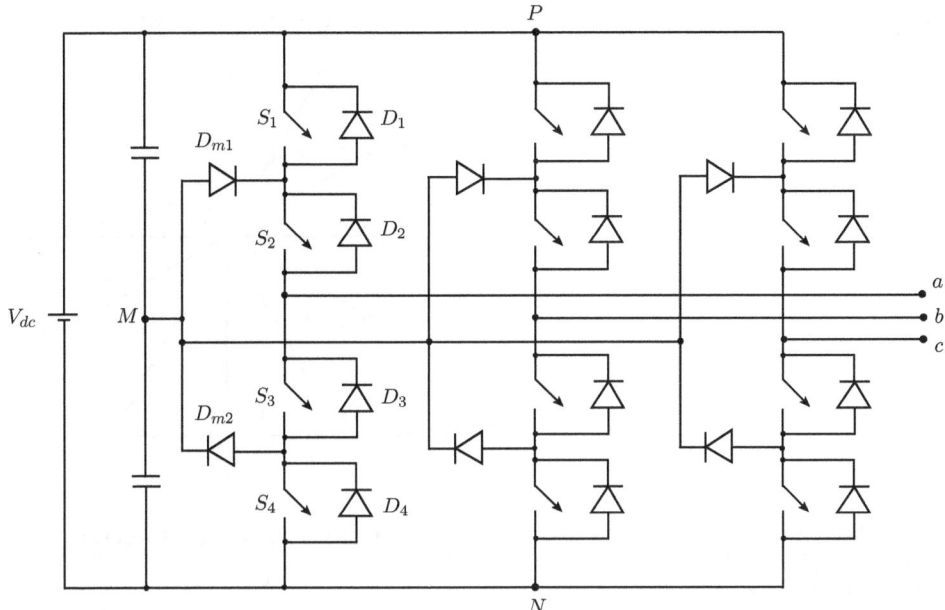

Fig. 6.4 Three-phase three-level converter

an exercise. From Fig. 6.4, it is clear that a three-phase converter merely involves replicating three converter legs.

For a three-phase converter, similar to the previous chapter, the total number of conduction modes is merely the number of conduction modes for a single converter leg raised to the power of the number of phases (3). This is due to the fact that each leg can be operated independently with respect to the other two legs without any danger of short-circuiting any electrolytic capacitors. Therefore, for a three-phase 3-level converter, the total number of conduction modes is:

$$3^3 = 3 \times 3 \times 3 = 27 \qquad (6.1)$$

And for a three-phase 4-level converter, the total number of conduction modes is:

$$4^3 = 4 \times 4 \times 4 = 64 \qquad (6.2)$$

For each conduction mode, one can calculate the line-line output voltages v_{ab}, v_{bc}, and v_{ca} or line-neutral output voltages v_{an}, v_{bn} and v_{cn} depending on the requirement of the load or ac system to which the converter is connected. Subsequently, the output voltages can be transformed using Clarke's transformation into the complex α-β domain and represented as voltage vectors [8].

Due to the large number of conduction modes, it would simpler to present the results through Python code. The reader can refer to the previous chapter for details on how this

Fig. 6.5 Vectors in the complex α-β plane 3-phase 3-level converter

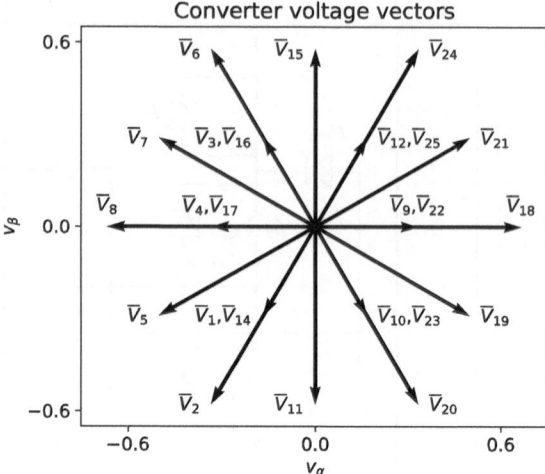

Fig. 6.6 Vectors in the complex α-β plane 3-phase 4-level converter

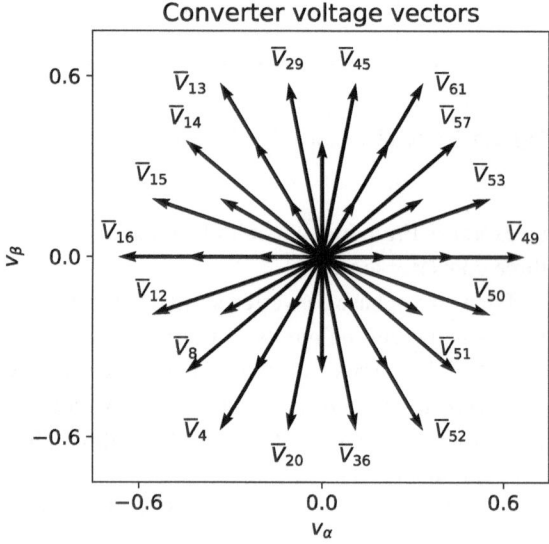

process can be performed in a step-by-step manner. The Python code can be found in the file line_neutral_map_3level.py inside folder vector_diagrams within the folder chapter6 at the link: https://github.com/opensourceelectrical/switching-strategies-for-power-electronics for the three-phase 3-level converter, and in file line_neutral_map_4level.py inside folder vector_diagrams within the folder chapter6 at the link: https://github.com/opensourceelectrical/switching-strategies-for-power-electronics for the three-phase 4-level converter. Table 6.3 shows the switching combinations for the voltage vectors of the three-phase 3-level converter. Figures 6.5 and 6.6 show the converter voltage vector diagrams for the three-phase 3-level converter and the three-phase 4-level converter

Table 6.3 3-level converter voltage vectors

Phase a				Phase b				Phase c				Vector
S_1	S_2	S_3	S_4	S_1	S_2	S_3	S_4	S_1	S_2	S_3	S_4	
0	0	1	1	0	0	1	1	0	0	1	1	$\overline{V}_0$
0	0	1	1	0	0	1	1	0	1	1	0	$\overline{V}_1$
0	0	1	1	0	0	1	1	1	1	0	0	$\overline{V}_2$
0	0	1	1	0	1	1	0	0	0	1	1	$\overline{V}_3$
0	0	1	1	0	1	1	0	0	1	1	0	$\overline{V}_4$
0	0	1	1	0	1	1	0	1	1	0	0	$\overline{V}_5$
0	0	1	1	1	1	0	0	0	0	1	1	$\overline{V}_6$
0	0	1	1	1	1	0	0	0	1	1	0	$\overline{V}_7$
0	0	1	1	1	1	0	0	1	1	0	0	$\overline{V}_8$
0	1	1	0	0	0	1	1	0	0	1	1	$\overline{V}_9$
0	1	1	0	0	0	1	1	0	1	1	0	$\overline{V}_{10}$
0	1	1	0	0	0	1	1	1	1	0	0	$\overline{V}_{11}$
0	1	1	0	0	1	1	0	0	0	1	1	$\overline{V}_{12}$
0	1	1	0	0	1	1	0	0	1	1	0	$\overline{V}_{13}$
0	1	1	0	0	1	1	0	1	1	0	0	$\overline{V}_{14}$
0	1	1	0	1	1	0	0	0	0	1	1	$\overline{V}_{15}$
0	1	1	0	1	1	0	0	0	1	1	0	$\overline{V}_{16}$
0	1	1	0	1	1	0	0	1	1	0	0	$\overline{V}_{17}$
1	1	0	0	0	0	1	1	0	0	1	1	$\overline{V}_{18}$
1	1	0	0	0	0	1	1	0	1	1	0	$\overline{V}_{19}$
1	1	0	0	0	0	1	1	1	1	0	0	$\overline{V}_{20}$
1	1	0	0	0	1	1	0	0	0	1	1	$\overline{V}_{21}$
1	1	0	0	0	1	1	0	0	1	1	0	$\overline{V}_{22}$
1	1	0	0	0	1	1	0	1	1	0	0	$\overline{V}_{23}$
1	1	0	0	1	1	0	0	0	0	1	1	$\overline{V}_{24}$
1	1	0	0	1	1	0	0	0	1	1	0	$\overline{V}_{25}$
1	1	0	0	1	1	0	0	1	1	0	0	$\overline{V}_{26}$

respectively. The reader is encouraged to verify the vector diagram of Fig. 6.5 with respect to the switching combinations of Table 6.3. For the 4-level converter, due to the large number of vectors, only the largest outermost vectors have been labelled. The reader can generate the complete vector diagram by executing the Python code. Both vector diagrams appear extremely complex due to the large number of vectors, and at first glance, it may appear that the logic for deciding the vectors would be extremely complex. But, if one divides up the vector diagram into quadrants, the logic can be formulated in a systematic manner as will be described in the next section.

From Figs. 6.5 and 6.6, a few observations can be made, though it might need one to take a closer look. The mutli-level nature of the converters is very clear from the fact that the converter voltage vectors have different magnitudes and seemed to be layered. In the case of the 3-level vector map of Fig. 6.5, there are three zero vectors, twelve inner vectors with the same magnitude as in the 2-level converter described in the previous chapter and twelve outer vectors that are new and the result of the higher voltage dc bus. In the case of the 4-level vector map of Fig. 6.6, there are four zero vectors, 18 inner vectors similar to the 2-level vector map, 24 intermediate vectors similar to the outer vectors of the 3-level vector map and finally 18 outermost vectors which are the result of the higher voltage dc bus. Therefore, a multi-level converter merely adds layers of converter voltage vectors over the converter voltage vectors produced by lower level converters.

In the case of the voltage vectors of the 3-level converter shown in Fig. 6.5, the inner voltage vectors have a redundancy. Each vector can be realized through two switching combinations. In the case of the voltage vectors of the 4-level converter shown in Fig. 6.6, each of the innermost voltage vectors can be achieved through three switching combinations, while the each of the intermediate vectors can be achieved through two switching combinations. At first, this might seem like a source of confusion for choosing the actual vector in a sequence during a switching cycle. However, this redundancy is of great help in balancing the voltages of the electrolytic capacitors that form the dc bus. Though two converter voltage vectors produce identical three-phase output voltages, one might cause a particular capacitor to charge while another might cause another capacitor to discharge. Though capacitor voltage balancing is extremely important in the operation of a multi-level converter, it is fairly complex and quite often closely associated with the closed-loop control scheme in use. Therefore, though a brief mention will be made of this aspect in the next section, capacitor voltage balancing will not be covered in great detail while describing the switching strategy.

This section has described the basic philosophy of operation of a multi-level converter. Though there are similarities with that of the 2-level converter described in the previous chapter, due to the larger number of devices in a converter leg, multiple devices need to be turned ON to produce a certain output voltage. Furthermore, the forbidden conduction modes consist of all those conduction modes that result in the short-circuit of either the entire dc bus or one or more capacitors of the dc bus. Since multi-level converters are best suited for high power applications which are either three-phase or multi-phase, Clarke's transformation can be applied to the output voltages produced by each and every switching combination to produce a vector diagram. Due to the large number of switching combinations, these vector maps are generated completely using Python code, though the reader is encouraged to review the description of the process in the previous chapter. The next section will describe the algorithm for choosing converter voltage vectors for a required three-phase output voltage.

6.5   SVPWM for Multi-level Converters

The previous section described the operation of multi-level converter legs with the examples of the 3-level and the 4-level converter legs. Following the usual method of defining forbidden conduction modes and subsequently, the allowable conduction modes in order to produce an output voltage, the output voltages of three-phase converters built using these multi-level converters can be computed using Python code. Using the same procedure as in the previous chapter, the output voltages of the three-phase converter can be transformed using Clarke's transformation and a vector diagram is produced using all possible switching combinations of the converters. Though the vector diagrams for multi-level converters were far more complicated with a far greater number of voltage vectors, it was possible to draw a few inferences based on these vector diagrams. This section will dive deeper into these vector diagrams with the objective of formulating an algorithm for implementing SVPWM for multi-level converters.

Though it might be of interest to some to formulate a generalized algorithm for any k-level converter, in this section, the focus will be on formulating algorithms specifically for the 3-level converter whose vector diagram is shown in Fig. 6.5 [9]. However, as already described in the previous section, the vector map of the 4-level converter is a mere extension of the 3-level converter, due to which the algorithm for SVPWM for the 4-level converter will be an extension of the algorithm for SVPWM for the 3-level converter. The procedure followed will be quite similar to that in the previous chapter and will heavily use basic geometry and trigonometric expressions. Therefore, to fully follow the discussion in this section, the reader would need to think more like a high-school student rather than a power electronics engineer.

Let us start this discussion with the vector diagram of the 3-level converter shown in Fig. 6.5. Due to the larger number of vectors, it would be advisable to focus on a segment of the vector diagram rather than the entire vector diagram. If one observes the vector diagram of Fig. 6.5, it is clear that there is a symmetry over each quadrant. Therefore, if we focus on the part of the vector diagram where $v_\alpha > 0$ and $v_\beta > 0$, the segment in Fig. 6.7 is produced. The Python code for this vector diagram can be found in file `three_level_one_quadrant.py` inside folder `vector_diagrams` within the folder `chapter6` at the link: https://github.com/opensourceelectrical/switching-strategies-for-power-electronics.

Along with the converter voltage vectors, dashed lines joining the vector tips can be seen in Fig. 6.7 which result in a number of triangles. Since it is quite easy to get visually confused and think of several triangles, we must focus on those triangles that result in optimized switching. To be able to limit ourselves to only those triangles that result in optimized switching, we can draw a few inferences from SVPWM implemented on a 2-level converter as described in the previous chapter.

In Fig. 6.7, two circular trajectories have been plotted for possible required output voltage vectors. The inner trajectory which is voltage vector of magnitude 0.2 is well within the inner

Fig. 6.7 Vector diagram for
3-phase 3-level converter—one
quadrant of α-β plane

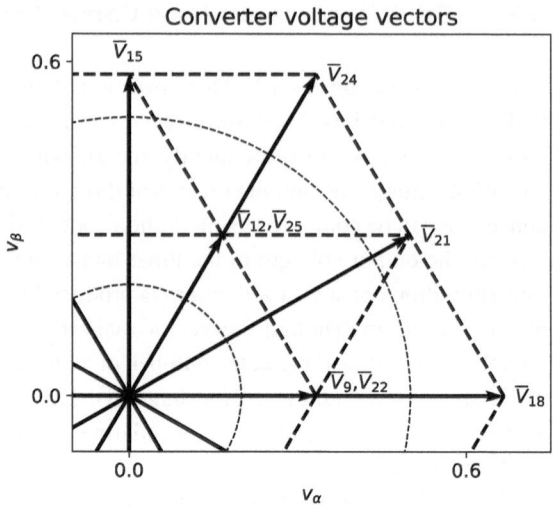

hexagon that is similar to that of the 2-level converter described in the previous chapter. Therefore, in order to produce output voltages corresponding to the inner output voltage vector trajectory, one needs only the vectors on the inner hexagon and therefore, the algorithm used for selecting the vectors will be a minor modification of the algorithm presented in the previous chapter. However, the inner hexagon has an important property that we must look for in the outer hexagon to be able to extend SVPWM for a higher level converter. The inner hexagon can be divided into six equilateral triangles, and each triangle is comprised of two converter voltage vectors. To achieve optimized switching that is both symmetric and also results in the minimum number of switching transitions, these two converter voltage vectors are used in combination with the zero voltage vectors that are at the origin of the hexagon. One could state that the three corners of the triangle are formed by converter voltage vectors, two of them being non-zero voltage vectors while the third being the zero vector. The reader is encouraged to review the discussion in the previous chapter, and also verify from Fig. 6.5, that for the inner hexagon, combining the inner converter voltage vectors with the zero vectors will always result in the minimum number of switching transitions in a switching cycle.

We must extend this logic to the case when the trajectory of the required output voltage vector lies outside the inner hexagon. This is shown in Fig. 6.7 with a circular trajectory corresponding to a voltage vector with a magnitude of 0.5. Clearly, this outer vector trajectory lies completely outside the inner hexagon. The reader is encouraged to review the simulation results of the previous chapter, where the vector diagram was used to determine the capacity of the converter. Therefore, the output voltages corresponding to this vector trajectory cannot be produced by using the vectors of the inner hexagon, and it is essential to also use the outer vectors of the multi-level converter. We must choose the minimum number of converter voltage vectors to be able to realize a particular required output voltage vector. Furthermore, the converter voltage vectors should be chosen such that the minimum number of switching transitions occur in a switching cycle.

The reader is encouraged to examine closely the vectors in the quadrant of Fig. 6.7 with the switching combinations of Table 6.3. Let us first examine the equilateral triangles that can be seen to form the outer layer of the vector diagram. These equilateral triangles are comprised of converter voltage vectors that differ from one another by a single switching transition. Furthermore, one can observe the minimum possible change in the converter output voltage vector as one moves between the vectors that form the vertices of the triangles. As an example, let us consider the equilateral triangle formed by the converter voltage vectors $\overline{V}_9$ (or $\overline{V}_{22}$), $\overline{V}_{18}$ and $\overline{V}_{21}$. From Table 6.3, $\overline{V}_9$ differs from $\overline{V}_{18}$ by a single device transition (one device turning OFF and another device turning ON), and $\overline{V}_{18}$ differs from $\overline{V}_{21}$ by a single device transition.

Figure 6.8 shows triangles that are not equilateral triangles. Let us now examine a triangle that is not an equilateral triangle such as the one formed by the converter voltage vectors $\overline{V}_{13}$ (zero vector not shown), $\overline{V}_{12}$ (or $\overline{V}_{25}$), $\overline{V}_{21}$ and $\overline{V}_{26}$ (zero vector not shown). The reader is encouraged to synthesize a sequence using these converter voltage vectors. In order to maintain a single switching transition between the converter voltage vectors, it would be necessary to choose a sequence such as $\overline{V}_{13}$–$\overline{V}_{12}$–$\overline{V}_{21}$–$\overline{V}_{25}$–$\overline{V}_{26}$. However, quite clearly, this results in the need to choose five converter voltage vectors in a single switching cycle, and therefore, will increase the total number of switching transitions in the switching cycle. Instead of choosing this non-optimal triangle, we could choose either the equilateral triangle formed by $\overline{V}_0$, $\overline{V}_9$, $\overline{V}_{12}$ and $\overline{V}_{13}$ in the inner hexagon, or the equilateral triangle formed by $\overline{V}_9$ (or $\overline{V}_{22}$), $\overline{V}_{12}$ (or $\overline{V}_{25}$) and $\overline{V}_{21}$ in outer hexagon. We are already aware that the vectors in the inner hexagon satisfy both criteria—minimum number of converter voltage vectors and minimum number of switching transitions. Let us examine the equilateral triangle of the outer hexagon.

Fig. 6.8 Vector diagram for 3-phase 3-level converter—mixed inner and outer voltage vectors

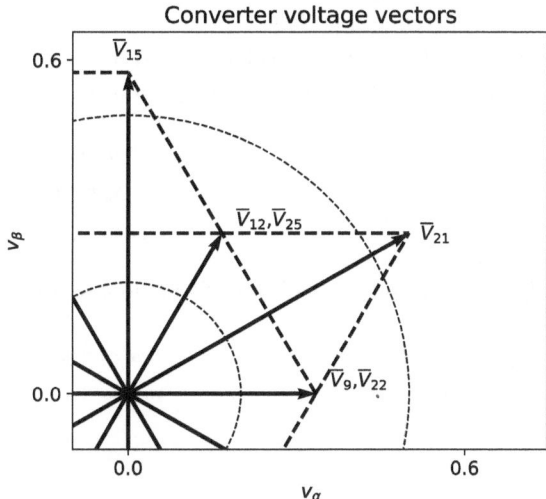

In the case of the equilateral triangle formed by the converter voltage vectors $\overline{V}_9$ (or $\overline{V}_{22}$), $\overline{V}_{12}$ (or $\overline{V}_{25}$) and $\overline{V}_{21}$, there are two possibilities. The first is if we choose the vector sequence to be $\overline{V}_9$–$\overline{V}_{12}$–$\overline{V}_{21}$–$\overline{V}_{22}$. The reader is encouraged to verify that the number of switching transitions will be minimum as each vector differs from the previous in the sequence by a single switching transition. The only difference with respect to a triangle in the inner hexagon is that the sequence does not start with a zero vector. However, this is only expected, as to realize a larger output voltage vector, one would need to choose larger converter voltage vectors. However, despite choosing a non-zero vector to be the first in the sequence, the first and last converter voltage vectors in the sequence ($\overline{V}_9$ and $\overline{V}_{22}$) have identical v_α and v_β values. The second possibility is to choose the vector sequence to be $\overline{V}_{25}$–$\overline{V}_{22}$–$\overline{V}_{21}$–$\overline{V}_{12}$. The reader is encouraged to verify that this sequence will also satisfy both criteria and is also a valid choice. To be able to choose between these two vector sequences, one needs to know the voltage across the capacitors that form the dc bus. Even though two voltage vectors produce identical output voltages, due to the different switching combinations, the choice of converter voltage vectors can cause capacitors to charge or discharge in a different manner, and therefore, can be used to equalize the voltage of the dc bus capacitors.

To be able to deduce at an abstract level, a logic for SVPWM for multi-level converters, we must divide the vector diagram into segments that result in the least number of converter voltage vectors in a sequence, and that results in the minimum number of switching transitions per switching cycle. The choice of equilateral triangles guarantees a selection of converter voltages that satisfies these two criteria. On the other hand, choosing other types of triangles results in converter voltage vectors that differ from each other to a greater extent. Therefore, to ensure that the minimum number of switching transitions occur in a switching cycle, one is required to select a larger number of converter voltage vectors to form the vector sequence. As the required output voltage vector follows the circular trajectory as shown in Fig. 6.7, we must formulate an algorithm that selects these equilateral triangles. For higher-level converters, one will find different types of triangles besides the optimal equilateral triangles, and the reader is encouraged to plot these for the 4-level converter vector diagram of Fig. 6.6.

In order to formulate an algorithm to choose converter voltages for particular three-phase output voltages, let us first begin by choosing a generic three-phase voltage template:

$$
\begin{aligned}
v_{ra} &= V_m \cos(\omega t) \\
v_{rb} &= V_m \cos(\omega t - 120°) \\
v_{rc} &= V_m \cos(\omega t - 240°)
\end{aligned}
\tag{6.3}
$$

Upon performing Clarke's transformation, we obtain:

$$
\begin{aligned}
v_{r\alpha} &= V_m \cos(\omega t) \\
v_{r\beta} &= V_m \sin(\omega t)
\end{aligned}
\tag{6.4}
$$

Since we wish to superimpose this $v_{r\alpha}$ and $v_{r\beta}$ in the form of an output voltage vector on the vector diagram of Fig. 6.7, we must define the range of the peak V_m of the output voltages.

The reader can refer to the previous chapter which described how the capacity of a three-phase converter can be determined using the converter voltage vector diagram. In the case of the 3-level converter, the capacity is easier to determine as it is directly equal to the magnitude of the vectors $\overline{V}_5$, $\overline{V}_7$, $\overline{V}_{15}$, $\overline{V}_{19}$ and $\overline{V}_{21}$. From the Python code, this can be found to be 0.5773 assuming $V_{dc} = 1$, or to be generic, can be considered to be 0.5773 V_{dc}. At any given instant of time t_1, one can imagine any three-phase output voltage having a magnitude from 0 to 0.5773 and at some angle $\omega t_1 = \theta_1$ with respect to the α axis. For the purpose of describing how an algorithm can be formulated, let us assume that $v_{r\alpha 1} > 0$ and $v_{r\beta 1} > 0$ and so the quadrant selected in Fig. 6.7 is sufficient for determining the converter voltage vectors. One can then alter this algorithm for the remaining three quadrants.

The entire algorithm for the quadrant is shown in Figs. 6.9, 6.10 and 6.11. A systematic approach to formulating an algorithm is to gradually move from left to right in the vector diagram of Fig. 6.7 and determine the conditions for which the output voltage vector will be in a particular triangle. As stated before, the approach followed requires merely high school trigonometry and geometry, and the reader is strongly encouraged to use a pencil and paper

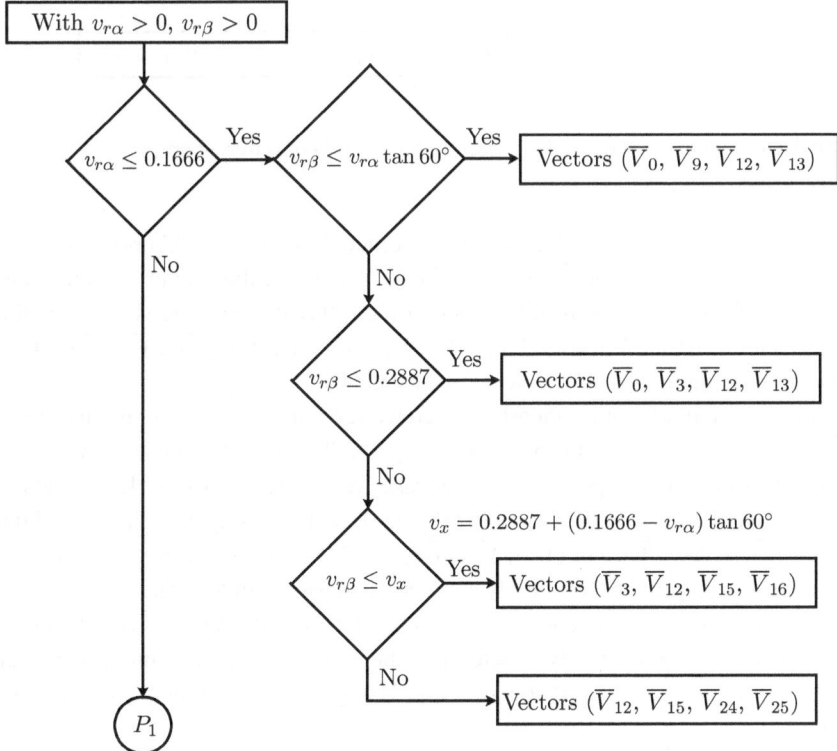

Fig. 6.9 Determining converter voltage vectors for 3-level converter—part 1

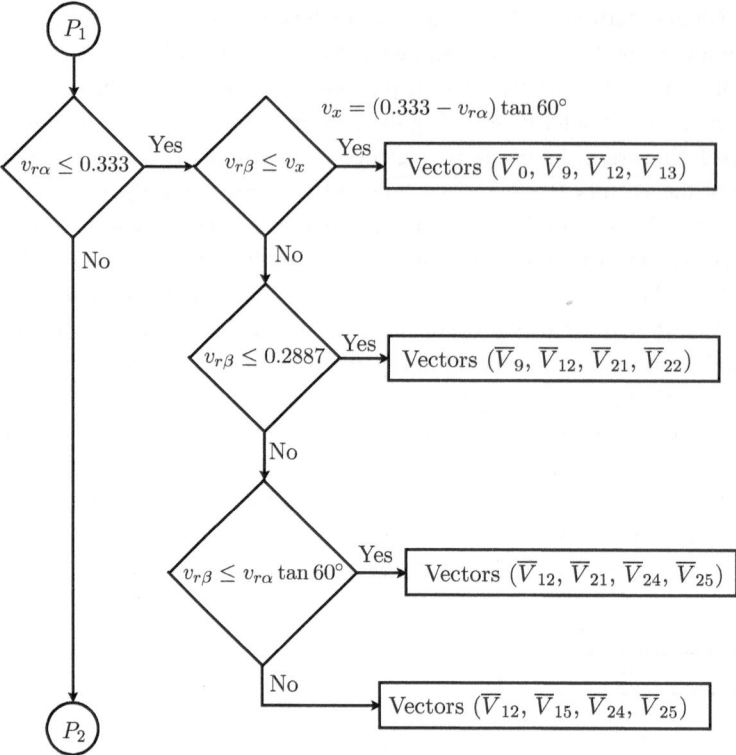

Fig. 6.10 Determining converter voltage vectors for 3-level converter—part 2

to follow along. To begin with let us draw a vertical line at $v_\alpha = 0.1666$ which will pass through the tips of the vectors $\overline{V}_{12}, \overline{V}_{25}$. If $v_{r\alpha 1} < 0.1666$, the output voltage vector will lie to the left of this line. This results in one of the following possible vector combinations $(\overline{V}_0, \overline{V}_9, \overline{V}_{12}, \overline{V}_{13})$, $(\overline{V}_0, \overline{V}_3, \overline{V}_{12}, \overline{V}_{13})$, $(\overline{V}_3, \overline{V}_{12}, \overline{V}_{15}, \overline{V}_{16})$ or $(\overline{V}_{12}, \overline{V}_{15}, \overline{V}_{24}, \overline{V}_{25})$. The exact vector combination will be determined by the value of $v_{r\beta 1}$.

If $v_{r\beta 1} \leq v_{\alpha 1} \tan 60°$, quite clearly the converter voltage vector combination is $(\overline{V}_0, \overline{V}_9, \overline{V}_{12}, \overline{V}_{13})$. If $v_{r\beta 1} > v_{\alpha 1} \tan 60°$ but $v_{r\beta 1} \leq 0.2887$, then the converter voltage vector combination is $(\overline{V}_0, \overline{V}_3, \overline{V}_{12}, \overline{V}_{13})$. In this case we are merely using the v_β value of the converter voltage vector $\overline{V}_{12}$. If $v_{r\beta 1} > 0.2887$, we have remaining two vector combinations, namely $(\overline{V}_3, \overline{V}_{12}, \overline{V}_{15}, \overline{V}_{16})$ or $(\overline{V}_{12}, \overline{V}_{15}, \overline{V}_{24}, \overline{V}_{25})$. However, this is where the condition becomes a little tricky and needs merely geometry and trigonometry. The dashed line joining converter voltage vectors $\overline{V}_{12}$ and $\overline{V}_{15}$ makes an angle of 60° degrees with the horizontal. However to use this angle directly to determine the limit of $v_{r\beta 1}$, we cannot use $v_{r\alpha 1}$ directly as before, but rather must use $0.1666 - v_{r\alpha 1}$. Using this line segment, we can apply the condition:

$$v_{r\beta 1} \leq 0.2887 + (0.1666 - v_{r\alpha 1}) \tan 60° \tag{6.5}$$

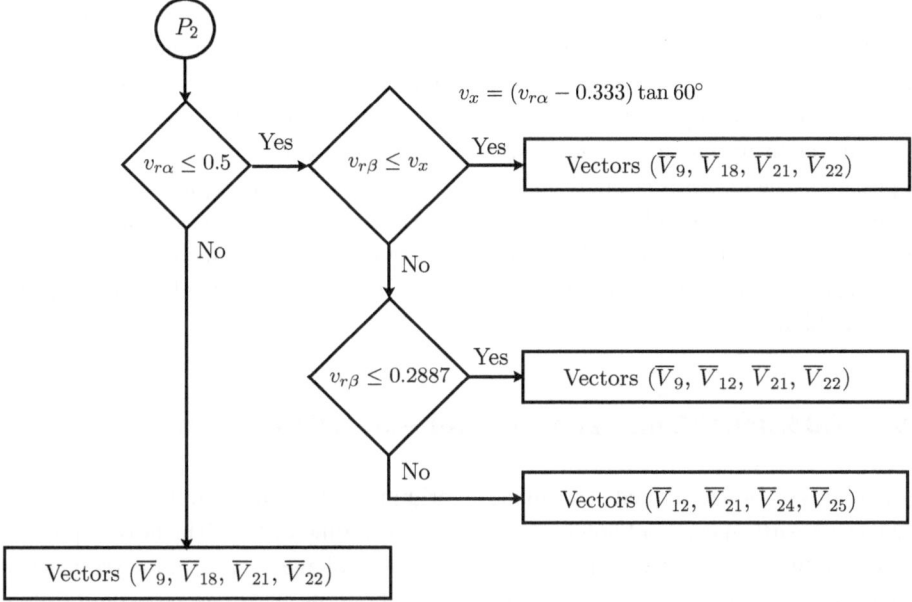

Fig. 6.11 Determining converter voltage vectors for 3-level converter—part 3

Under this condition, the converter vector combination is $(\overline{V}_3, \overline{V}_{12}, \overline{V}_{15}, \overline{V}_{16})$. And if this condition also fails, the last choice that remains is $(\overline{V}_{12}, \overline{V}_{15}, \overline{V}_{24}, \overline{V}_{25})$. Using this same approach, one can divide the quadrant up using vertical lines at $v_\alpha = 0.3333$ and $v_\alpha = 0.5$ and use similar conditional arguments to move upwards from one triangle to another.

Now that we have developed the algorithm for the first quadrant, the approach can be replicated for the remaining three quadrants. The readers are strongly encouraged to attempt to draw similar flowcharts for the remaining three quadrants. The logic can be found in the simulation of the 3-level 3-phase converter using SVPWM in file SVPWM.py inside folder three_level_converter within the folder chapter6 at the link: https://github.com/opensourceelectrical/switching-strategies-for-power-electronics.

Moreover, now that the logic has been developed for the 3-level converter, this logic can be further extended for the 4-level 3-phase converter. The reader is also encouraged to examine ways in which the determination of the voltage vector combination can be accelerated. As an example, in the algorithm described, we progressed from the interior of the quadrant to the exterior. However, if it is known that the required output voltage vector has a magnitude close to the maximum capacity of the converter, one can choose the reverse approach in our conditional checks, i.e to progress from the exterior to the interior of the quadrant. This can potentially save a few conditional checks and result in faster vector determination.

This section described how the algorithm for SVPWM for a multi-level converter can be implemented in a systematic manner despite the very large number of converter voltage vectors that are inevitable in any multi-level topology. Though the algorithm seems tedious, a quick observation will yield that most computations are simple arithmetic that can be performed using even basic microcontrollers. Moreover, it takes a few conditional checks to determine which voltage vector combination should be chosen which makes the SVPWM algorithm relatively fast to execute despite the large number of vector combinations that are possible. The reader is strongly encouraged to write the entire logic for the 3-level and 4-level converters before referring to the SVPWM code provided in the simulation package. The next section will describe a few changes that are necessary in the computation of the time intervals for the converter voltage vectors, as well how they should be chosen in the case of ambiguity.

6.6   Additional Constraints for Multi-level Converters

The previous section described in detail the algorithm for choosing converter voltage vectors to achieve a particular output voltage vector at a given time instant. The next step remains to compute the time intervals for which the converter voltage vectors in the sequence will be applied. Though one might be tempted to merely apply the same procedure followed for 3-phase 2-level converters in the previous chapter, one needs to take into account a few differences in the vector diagram of multi-level converters with respect to their 2-level counterparts. Additionally, the previous section described how while choosing converter voltage vectors in a sequence, there are at times multiple choices. This section will deal with these two issues - the computation of the time intervals of the converter voltage vectors, and the manner in which the converter voltage vectors can be chosen when there is ambiguity.

To begin the discussion on calculating the time intervals for which converter voltage vectors must be applied, let us repeat the basic equation for 2-level converters from the previous chapter:

$$\overline{V}_{ref} T = \overline{V}_x t_1 + \overline{V}_y t_2 \tag{6.6}$$

In the above equation, $\overline{V}_{ref}$ is the instantaneous value of the required output voltage vector, and $\overline{V}_x, \overline{V}_y$ are the non-zero converter voltages in the triangle in which the output voltage vector is located. In addition to $\overline{V}_x, \overline{V}_y$, zero voltage vectors $\overline{V}_0$ and $\overline{V}_7$ will also be applied to complete the sequence, and the time interval of the zero vectors is expressed as:

$$t_0 = T - (t_1 + t_2) \tag{6.7}$$

The above expressions will also hold true for multi-level converters if the required output voltage vector lies in any of the innermost triangles. The only difference will be in the choice of the zero vector, as the zero vectors are chosen to minimize the number of switching transitions.

If the required output voltage vector lies in a triangle outside the innermost triangles, the voltage vector equation changes as:

$$\overline{V}_{ref}T = \overline{V}_x t_1 + \overline{V}_y t_2 + \overline{V}_z t_0 \tag{6.8}$$

This is due to the fact that in an outer triangle, the zero vectors are not chosen any more, and all the vectors that form the vertices of the triangle are non-zero vectors. If the reader reviews the algorithms in the previous chapter, and compares that with the vector diagrams of Fig. 6.7, there will not be a single vector $\overline{V}_z$, but rather two vectors $\overline{V}_{z1}$ and $\overline{V}_{z2}$ which have identical values, due to which a single value of $\overline{V}_z$ in the above expression is correct. The reader is encouraged to review this fact as it is extremely important for the discussion that follows. In the previous chapter, we had separated the above equation into real and imaginary parts, and solved them simultaneously for the variables t_1 and t_2. However, we now have two equations in three variables, namely t_0, t_1 and t_2.

The above issue can be solved in two ways. We can add the following third equation:

$$t_0 + t_1 + t_2 = T \tag{6.9}$$

With this, we have three variables and three equations if we combine the real and imaginary parts of the above equation, and this system of equations can be solved either in a matrix form or by substitution. Since, the calculation of the time intervals need to be performed in real-time in a microcontroller, it would be best if these equations are simplified into a form similar to the expressions for t_1 and t_2 presented in the previous chapter. This can be achieved through substitution as follows:

$$\overline{V}_{ref}T = \overline{V}_x t_1 + \overline{V}_y t_2 + \overline{V}_z(T - t_1 - t_2) \tag{6.10}$$

It is possible to expand the above expression into real and imaginary parts and simplify them to result in final expressions for t_1 and t_2 which can be calculated as a simple arithmetic expression in a microcontroller. However, we could instead use a neat trick to continue using the expressions for t_1 and t_2 in the previous chapter. The issue we are facing is the introduction of the non-zero vector $\overline{V}_z$. If we however, define new vectors as follows:

$$\begin{aligned}
\overline{V}'_{ref} &= \overline{V}_{ref} - \overline{V}_z \\
\overline{V}'_x &= \overline{V}_x - \overline{V}_z \\
\overline{V}'_y &= \overline{V}_y - \overline{V}_z \\
\overline{V}'_z &= \overline{V}_z - \overline{V}_z
\end{aligned} \tag{6.11}$$

Among these, the last vector $\overline{V}'_z$ is very obviously zero and therefore a zero vector. With the above definitions, we have in essence "shifted" the triangle to the origin of the vector diagram with one of the vertices now being at zero. The question one might ask is, can such an operation be legal?

To answer this question, we merely have to rewrite the expression with these new vectors:

$$\overline{V'}_{ref}T = \overline{V'}_x t_1 + \overline{V'}_y t_2 + \overline{V'}_z t_0$$

$$(\overline{V}_{ref} - \overline{V}_z)T = (\overline{V}_x - \overline{V}_z)t_1 + (\overline{V}_y - \overline{V}_z)t_2 + (\overline{V}_z - \overline{V}_z)t_0$$

$$\overline{V}_{ref}T - \overline{V}_z T = \overline{V}_x t_1 + \overline{V}_y t_2 + \overline{V}_z t_0 - \overline{V}_z(t_1 + t_2 + t_0) \qquad (6.12)$$

$$\overline{V}_{ref}T = \overline{V}_x t_1 + \overline{V}_y t_2 + \overline{V}_z t_0$$

Therefore, we have not altered the original expression. However, due to the fact that $\overline{V'}_z$ is a zero vector, the expression is simplified to:

$$\overline{V'}_{ref}T = \overline{V'}_x t_1 + \overline{V'}_y t_2 \qquad (6.13)$$

Subsequently, one can separate the above expression into real and imaginary parts and produce similar expressions for t_1 and t_2. These expressions are repeated here for convenience:

$$t_1 = \frac{v'_{r\alpha} v'_{y\beta} - v'_{y\alpha} v'_{r\beta}}{v'_{x\alpha} v'_{y'\beta} - v'_{x\beta} v'_{y\alpha}} T$$

$$t_2 = \frac{v'_{r\beta} v'_{x\alpha} - v'_{x\beta} v'_{r\alpha}}{v'_{x\alpha} v'_{y\beta} - v'_{x\beta} v'_{y\alpha}} T \qquad (6.14)$$

One might argue that we are now performing more computations with the calculation of these new vectors. However, this is a minor arithmetic computation, and furthermore the choice of $\overline{V}_z$ is very clear as it is always the first vector (or the last vector as they are identical in value) in any sequence of vectors for a triangle. The reader is of course welcome to expand the Eq. (6.10) and solve the simultaneous equations to derive expressions for t_1 and t_2 and compare them with the expressions above. However, the above calculations can be used for every triangle in the vector diagram of a multi-level converter. For the inner triangles, the vector $\overline{V}_z$ will be the zero vector anyway, and therefore, one can either perform this trivial calculation, or can have a conditional check to verify whether $\overline{V}_z$ is zero before defining new variables, in case one wants to avoid any unnecessary calculations in real-time and reduce the burden for the microcontroller.

With this one issue resolved, we can deal with the issue of ambiguity in choosing vectors in a particular triangle. To begin with, this ambiguity usually exists in inner triangles and a few of the outer triangles. As an example, if we were to examine the triangle formed by $\overline{V}_9$ (or $\overline{V}_{22}$), $\overline{V}_{18}$ and $\overline{V}_{21}$, it is clear that the vector sequence can only be $\overline{V}_9 - \overline{V}_{18} - \overline{V}_{21} - \overline{V}_{22}$. The reader should refer to the description in the previous section that establishes the constraints for selecting the vectors in the sequence. First, consecutive vectors should differ by at most one switching transition. Second, the first and last vector on the sequence should have identical values and should differ only as switching combinations. One could choose the reverse sequence as well, namely $\overline{V}_{22} - \overline{V}_{21} - \overline{V}_{18} - \overline{V}_9$. However, the vectors in the sequence remain the same.

Let us now consider the example of $\overline{V}_9$ (or $\overline{V}_{22}$), $\overline{V}_{12}$ (or $\overline{V}_{25}$) and $\overline{V}_{21}$. In this case, one can think of two completely different vector sequences. The first could be $\overline{V}_9 - \overline{V}_{12} -$

$\overline{V}_{21}$–$\overline{V}_{22}$, while the other could be $\overline{V}_{25}$–$\overline{V}_{22}$–$\overline{V}_{21}$–$\overline{V}_{12}$. The reader is encouraged to verify that both of them are perfectly valid vector sequences as they adhere to the two conditions stated above. In such a case, which vector sequence should be selected if the required output voltage vector lies in the triangle defined by these converter voltage vectors? The answer to this lies in the state of the dc bus capacitor voltages. Though the two vector sequences will produce the same output voltage, the switching combinations that result in these vectors are very different. As an example, the vector $\overline{V}_9$ will require devices S_2–S_3 of phase a, S_3–S_4 of phase b and S_3–S_4 of phase c to be turned ON. It is very clear that this vector will only involve the lower dc bus capacitor. In comparison, the vector $\overline{V}_{25}$ will require devices S_1–S_2 of phase a, S_1–S_2 of phase b and S_2–S_3 of phase c to be turned ON. It is very clear that converter voltage vector $\overline{V}_{25}$ only involves the upper dc capacitor.

If we compare the vector sequences $\overline{V}_9$–$\overline{V}_{12}$–$\overline{V}_{21}$–$\overline{V}_{22}$ and $\overline{V}_{25}$–$\overline{V}_{22}$–$\overline{V}_{21}$–$\overline{V}_{12}$, the first sequence uses the lower capacitor to a greater extent while the second sequence uses the upper capacitor to a greater extent. Whether this will result in a particular capacitor charging or discharging will depend on the direction of current. If currents were to be leaving a leg terminal, the capacitor connected to the terminal will discharge as it will have to supply a current, while if the current were entering the terminal, the capacitor connected to the terminal would charge. If we were to only choose a particular set of vector sequences that involve only one capacitor (say the lower capacitor), one would find the voltages of the dc bus capacitors would begin to diverge with time. As the divergence in the capacitor voltage increases, with time the output voltage produced by the converter would become distorted and so will the currents. Therefore, it is essential that we choose the vector sequences so as to balance out the capacitor voltages.

As stated before in the book, closed loop control is a fairly vast domain and will not be covered in this book. The issue of capacitor voltage balancing in multi-level converters has been extensively researched over the past few decades and one can find many research papers that use various different control loops and algorithms. For the sake of completeness, a very basic algorithm will be presented here that achieved a very rudimentary form of capacitor voltage balancing. In the case of a 3-level converter, with only two dc bus capacitors, when one capacitor charges, the other will discharge. This is due to the fact the sum of their voltage will be determined the dc voltage source connected across them. We can therefore, define the difference between their voltages as:

$$\Delta v_c = v_{c1} - v_{c2} \tag{6.15}$$

This difference should ideally be zero. However, attempting to tightly regulate the two dc bus capacitor voltages to be equal without too much control action will need advanced control algorithms. Therefore, to keep the logic simple, let us define an upper value Δv_c^{max} and a lower value Δv_c^{min} that Δv_c should not exceed. Quite obviously, Δv_c^{max} will be a positive value and Δv_c^{min} will be a negative value. A convenient manner of choosing this maximum and minimum limit is as a percentage of the reference values of the capacitor voltage or the entire dc bus voltage. Therefore:

$$\Delta v_c^{max} = +k V_{dc}^{ref}$$
$$\Delta v_c^{min} = -k V_{dc}^{ref}$$

(6.16)

When the capacitor voltage difference Δv_c exceeds Δv_c^{max}, this implies the lower dc bus capacitor has discharged due to over-utilization in the converter voltage vectors being chosen. At this event, we need to change our logic in choosing converter voltage vectors, and wherever possible, use voltage vectors that utilize the upper dc bus capacitor. When this change in logic is made, throughout the vector selection algorithm for the entire vector diagram, wherever there exists an ambiguity, we must choose the vector sequence to be such that the upper capacitor is utilized to a greater extent. This will result in the upper dc bus capacitor voltage decreasing, and therefore, the lower dc bus capacitor increasing. Gradually, the capacitor voltage difference Δv_c will decrease to zero and then become negative. When the capacitor voltage difference Δv_c decreases below Δv_c^{min}, the upper capacitor has discharged and the lower capacitor has charged. We must again reverse the logic to ensure that throughout the algorithm, when choosing vector sequences, the voltage vectors should be chosen such that the lower capacitor is used to a greater extent [10]. This logic can be found in the simulation of the 3-level converter in file SVPWM.py inside folder three_level_converter within the folder chapter6 at the link: https://github.com/opensourceelectrical/switching-strategies-for-power-electronics.

This section described two issues that are specific to multi-level converters and were not faced while describing SVPWM for 2-level converters. However, as can be seen, with a very simple arithmetic manipulation, we can continue to use the same expressions for calculation of time intervals as used for 2-level converters in the previous chapter. Furthermore, in terms of choosing vector sequences when multiple sequences are possible, it is important to choose the one that achieves capacitor voltage balancing. For 4-level converters, the reader can verify that the inner triangles of the vector diagram have more than two choices of vector sequences. This is only natural as a 4-level converter has three dc bus capacitors and the vector sequences can be chosen to utilize a particular dc bus capacitor to a greater extent. The next section will present the simulation results of the 3-level converter.

6.7   Simulations

The previous sections described the implementation of SVPWM in a 3-phase 3-level converter with generalizations to higher level converters. The procedure was quite similar to that of the 3-phase 2-level converter described in the previous chapter, except for a few changes brought about by the larger number of vectors. Though the algorithm for choosing the converter vectors was similar, one needs to consider only the equilateral triangles with converter voltage vectors that are located in close proximity. Additionally, for the outer triangles, due to the use of only non-zero vectors in the vector sequence, the expressions for calculation of the time periods for which the converter voltage vectors are applied need to be modified. In

3-level and 4-level topologies, it was observed that many of the inner voltage vectors could be realized using multiple switching combinations, which in turn resulted in redundancies in the voltage vector sequences that could be chosen for a particular triangle. The choice of the exact voltage vector sequence will impact the charging and discharging of the dc bus capacitors, and to ensure that the dc bus capacitor voltage do not diverge, an algorithm was presented to choose the appropriate voltage vectors in a sequence. This section will present the simulation results based on these concepts.

The simulation of the 3-phase 3-level converter can be found in folder `three_level_converter` within the folder `chapter6` at the link: https://github.com/opensourceelectrical/switching-strategies-for-power-electronics.

Figures 6.12, 6.13 and 6.14 show the simulation results. Figure 6.12 shows the line-neutral three-phase output voltages produced by the converter. The multi-level nature of the converter is very clear from the number of levels in the each phase of the output. Figure 6.13

Fig. 6.12 3-phase 3-level converter line-neutral output voltages

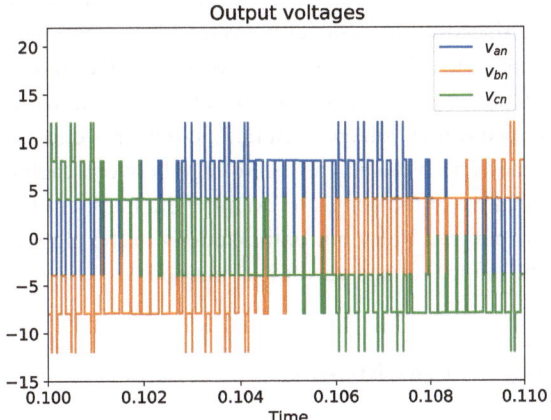

Fig. 6.13 3-phase 3-level converter output currents

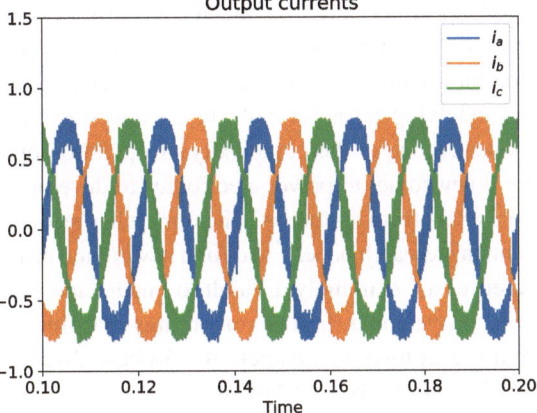

Fig. 6.14 3-phase 3-level converter dc bus capacitor voltages

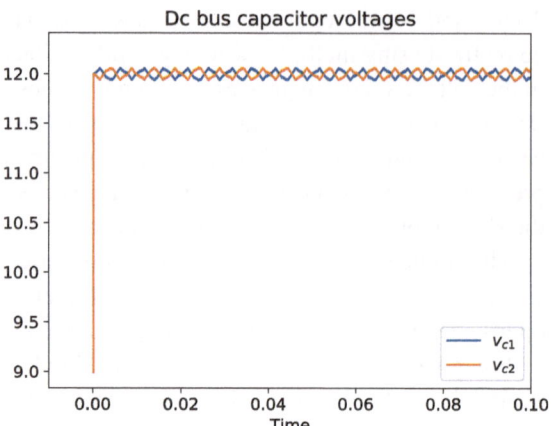

shows the output currents which can be seen to contain far less ripple that the output currents produced by the 2-level converter in the previous chapter. Figure 6.14 shows the dc bus capacitor voltages where it is very clear that though the capacitor voltages tend to diverge, the deviation in the capacitor voltages is limited to a hysteresis band due to the choice of the converter voltage vectors. The reader is encouraged to zoom into the plots and also plot other variables such as the voltage between each converter leg mid-point terminal and the negative of the dc bus to examine the number of levels being produced. The reader is also encouraged to try out optimizations in the algorithm for identifying the triangle in which the output voltage vector is contained.

6.8   Conclusions

This chapter has described in great detail the topology and operation of multi-level converters. Furthermore, the chapter has also described in great detail how SVPWM can be implemented for a 3-phase 3-level converter. A standard book on power electronics will present a number of different multi-level converter topologies such as cascaded converters and flying capacitor multi-level converters besides the NPC type multi-level converters. However, due to the complexity of operation of these converters, it is extremely difficult for a newcomer to power electronics to develop a reasonable understanding of multi-level converters from textbooks. The only other resource for learning state-of-the-art technology remains research papers, and for a newcomer, articles found in international journals or conferences are extremely difficult to understand due to the concise nature of these mediums. It is to overcome these shortcomings that this chapter delved into the depths of a single multi-level topology, namely the 3-phase 3-level NPC topology, with a few inferences to the 3-phase 4-level NPC topology.

The chapter describes how any k-level NPC converter leg can be constructed in terms of the number of dc bus capacitors, the number of controllable devices and their associated anti-parallel diodes, and finally the diodes that connect the intermediate nodes of the converter leg with the intermediate nodes of the dc bus. The chapter has also attempted to generalize how one can define the forbidden conduction modes for any k-level converter leg with specific references to the 3-level and 4-level converter legs. The chapter has also described how one can determine the allowable conduction modes for any multi-level converter leg. The reader is strongly encouraged to review the sections on multi-level converter topology and the operation of a multi-level converter leg, and extend the discussions in these sections to higher level converter legs such as the 5-level or the 7-level converter legs.

The chapter uses Python code to generate all the allowable conduction modes of the 3-phase 3-level converter and calculate the output voltage for each conduction mode. These output voltages are transformed using Clarke's transformation into converter voltage vectors that can be represented using a vector diagram. Readers are strongly encouraged to attempt to write the program on their own to fully understand how these vector diagrams come into being. With these vector diagrams, the chapter makes several observations about the operation of multi-level converters. These vector diagrams also leave open questions about several aspects of the converter voltage vectors. For example, what would be the impact of multiple converter voltage vectors being identical or how would one choose converter voltage vectors when so many of them are in close proximity? The chapter gradually addresses them while describing the process of SVPWM by defining the equilateral triangles within the vector diagram to be the fundamental structure within which the desired output voltage vector can be located.

The chapter has described in detail the algorithm for determining the triangle within which the instantaneous output voltage vector will be located. This algorithm is systematic and can be extended to higher level converters as well. Once the triangle is identified, the chapter describes how the sequence of converter voltage vectors can be selected to ensure minimum switching losses in a switching cycle while also maintaining symmetry of waveform over the switching cycle. The reader is strongly encouraged to formulate the algorithm for the entire vector diagram and attempt to translate it into Python code for simulation. The reader is also encouraged to attempt to extend the algorithm provided in the repository to a 3-phase 4-level converter using the same systematic approach to reach the outer triangles containing the larger converter voltage vectors.

The chapter describes how the expressions for calculating the time intervals of the converter voltages to be applied in a switching cycle can be used with minor modifications. Subsequently, the chapter describes the significance of multiple converter voltage vectors having the same identical value. Though multiple converter voltage vectors have the same value and are therefore located on the same vector in the diagram, they correspond to different switching combinations that may have very different conduction paths. These conduction paths have a very different impact on the charging and discharging cycles of the dc bus capacitors. Since the dc bus capacitors are connected in series, the dc voltage source applied across them maintains the sum of their voltages to be constant. However, the voltages of the individual dc bus capacitors can drift significantly resulting in distorted output voltages. To ensure that the voltages of the dc bus capacitors are approximately equal, the chapter describes a rudimentary strategy to ensure that different switching combinations are chosen. Though a practical multi-level converter might utilize advanced control strategies to balance out the dc bus capacitor voltages, the underlying philosophy remains the same—mix up the switching combinations so that all dc bus capacitors are equally utilized.

With this chapter, the contents of this book will come to an end. The next chapter will present the broad conclusions of this book and will also provide the reader with some pointers for future work and reading.

References

1. M. Malinowski, K. Gopakumar, J. Rodriguez, M.A. Perez, A survey on cascaded multilevel inverters. IEEE Trans. Ind. Electron. **57**(7), 2197–2206 (2010)
2. J. Rodriguez, S. Bernet, P.K. Steimer, I.E. Lizama, A survey on neutral-point-clamped inverters. IEEE Trans. Ind. Electron. **57**(7), 2219–2230 (2010)
3. M. Escalante, J.-C. Vannier, A. Arzande, Flying capacitor multilevel inverters and DTC motor drive applications. IEEE Trans. Ind. Electron. **49**(4), 809–815 (2002)
4. A. Stillwell, E. Candan, R.C.N. Pilawa-Podgurski, Active voltage balancing in flying capacitor multi-level converters with valley current detection and constant effective duty cycle control. IEEE Trans. Power Electron. **34**(11), 11 429–11 441 (2019)
5. J. Huang, K. Corzine, Extended operation of flying capacitor multilevel inverters. IEEE Trans. Power Electron. **21**(1), 140–147 (2006)
6. S. Du, A. Dekka, B. Wu, and N. Zargari, *Modular multilevel converters: analysis, control, and applications*. John Wiley & Sons, 2018
7. H.D. Tafti, A.I. Maswood, Advanced multilevel converters and applications in grid integration (2018)
8. A.K. Gupta, A.M. Khambadkone, A space vector PWM scheme for multilevel inverters based on two-level space vector PWM. IEEE Trans. Ind. Electron. **53**(5), 1631–1639 (2006)
9. V. Jayakumar, B. Chokkalingam, J.L. Munda, A comprehensive review on space vector modulation techniques for neutral point clamped multi-level inverters. IEEE Access **9**, 112 104–112 144 (2021)
10. S.A. Gonzalez, S.A. Verne, M.I. Valla, *Multilevel Converters for Industrial Applications* (CRC Press, 2013)

Conclusions

7

7.1 Summary of the Book

In this section, let us examine the learnings from every chapter. Since every chapter has its own section of conclusions, this section will instead highlight the lessons learned with respect to the book rather than with respect to each specific chapter, as the reader who has reached this chapter has read the book, and is in a position to reflect on the significance of it. The purpose of this section is to prompt the reader to stop and think, and potentially have another look at some of the chapters.

Chapter 1 introduces the book by describing the need for this book. The chapter describes how one of the greatest challenges faced by the power electronics industry is the lack of interest in young engineers to specialize in power electronics. The reason for the lack of interest is primarily the outdated mode of teaching and rather uninteresting educational content that continues to remain in the curriculum. The chapter describes how the IT industry revolutionized education for young engineers, making it possible for any young person interested in IT to be able to gain the skills necessary to enter the domain. The chapter describes how this book is a part of a larger project in making power electronics attractive to young engineers with the intention of "luring them in while they are young." Though such a phrase may appear vulgar and cheap, the chapter describes how most successful and established power engineers chose power electronics as their domain due to an early positive experience.

Chapter 2 introduced the concept of PWM with the example of either a single switch or a converter with a single power device that can be fed gating signals. The typical approach to teaching PWM to a newcomer to power electronics is describe the process of PWM merely with respect to the frequency response of the resultant waveform. However, rather than jumping directly to PWM in power electronics, the chapter examines the principle behind modulation. For this purpose, it uses modulation in the context of signal processing and communication, where an audio signal is mixed with a high frequency carrier signal.

© The Author(s), under exclusive license to Springer Nature Switzerland AG 2024
S. V. Iyer and M. N. Aalam, *Switching Strategies for Power Electronic Converters*,
Synthesis Lectures on Power Electronics, https://doi.org/10.1007/978-3-031-41405-3_7

The carrier signal by its name is a signal that "carries" the audio signal making it easier to transmit it over long distances. Using examples, it can be shown how the process of modulation results in a signal that still contains the audio signal, but in a completely different form, and which can be extracted at the receiving end using several techniques. In exactly the same manner, PWM mixes the reference that we would like the power converter to produce as an output, with a high frequency carrier signal. The result is a train of pulses of varying width that are used as gating signals to turn ON and turn OFF power devices in a power converter. Using frequency analysis, it can be observed that similar to modulation in communications, PWM in power electronics results in a waveform that contains the original signal. While in communications, it is necessary to demodulate the waveform to extract the audio signal, in power electronics, one merely needs to design a filter to eliminate those frequency components that are not needed.

Chapter 3 examined how the use of multiple power devices in a converter can provide additional features to a power converter, beginning with converters that use two controllable power devices. However, with multiple devices, one needs to co-ordinate the gating pulses provided to the devices, as for a particular feature of the converter, the devices would need to conduct in a certain manner. The chapter introduces the converter leg, which is a very popular structure comprising multiple controllable devices and diodes. These converter legs are available from many manufacturers and are also available in a number of different topologies for ready use in power converters. Using examples of buck, boost, buck-boost and half-bridge converters, the chapter describes how versatile this topology is, due to which it is considered a building block in power electronics.

Chapter 4 examines one of the most basic power converters that are used widely in industry, namely the full-bridge converter. The full-bridge converter is used in a number of different applications, for dc-dc power supplies, battery chargers, dc-ac converters for home and office UPS and many more. To understand the full-bridge converter, one merely needs to extend our learning from the converter leg presented in the previous chapter, and realize that the full-bridge converter is just two converter legs that can operate independently. The chapter describes a simple arithmetic approach to calculating the total number of allowable conduction modes of any power converter, based on the conduction states of the converter leg used as the building block, and the number of converter legs used to form the converter. Subsequent to the allowable conduction modes, one can determine the output voltage produced in each conduction mode. This allows for a vector representation of the converter output voltage, with a particular output vector corresponding to the switching combination. This vector representation though trivial for a full-bridge converter, makes it easy for the reader to understand how any converter can be represented as a variable voltage source—a concept that is used extensively in the later chapters.

Chapter 5 examines one of the most popular industrial power converters, namely the three-phase converter with only three output terminals for the three phases. Though three-phase converters differ vastly in topology depending on the three-phase connection, the topology presented in the chapter consisting of three converter legs is extremely powerful and applicable in a wide number of applications. The chapter extends the discussion on vector representation of converter output voltages by presenting a method to represent the output of the converter as a vector diagram using the very popular Clarke's transformation. Using the vectors in the vector diagram through a procedure known as SVPWM to precisely produce an output voltage, one can be assured of optimal utilization of the power converter. Due to this reason, many high power applications in three-phase systems use SVPWM in comparison to sine-triangle comparison based PWM. The chapter describes the algorithm in detail with a level of clarity which any newcomer to power electronics can understand.

Chapter 6 attempts to uncover a very complex topic in power electronics, namely high power converters. Due to the complexity of this topic, rather than providing an overview that would be difficult to digest, the chapter instead attempts to delve deep into a single fairly popular topology—the NPC multi-level converter. The chapter uses the SVPWM discussion from the previous chapter to describe in great detail how the vector diagram for a multi-level converter can be realized. Subsequently, the chapter describes in detail the algorithm for choosing the voltage vectors and the time intervals for which they should be applied. To complete the discussion to result in a working simulation, the chapter also describes how capacitor voltage balancing can be achieved in multi-level converters by incorporating an additional logic in the SVPWM algorithm. The chapter hopes to demystify an extremely complex domain of power electronics, but the knowledge of which is still very useful for an engineer hoping to enter the power industry.

7.2   Scope for Future Work

This section will describe ways in which an interested reader can extend the learnings from each chapter. Before diving into each chapter, one quick observation can be made of the entire book. The book avoided any closed loop control strategies due to the fact that control systems is a vast domain, and it would be impossible to cover it with any degree of detail while still attempting to dive into the details of switching strategies. For any interested reader with a background on control systems, one immediate suggestion will be to incorporate closed-loop control into every simulation presented in the book. Furthermore, the book avoided details of specific applications, as the focus was on describing generic switching strategies for practical power converters. For readers who wish to apply the contents of this book into their specific projects, it is highly encouraged to extend those specific simulations with additional control files that include models of appliances being used.

Chapter 2 described as an example a buck converter being one of the simplest power converters with a single controllable power device. The reader is encouraged to perform simulations of other basic power converters that use single controllable power devices such as boost converter, buck-boost converter, Cuk converter, Single-Ended Primary-Inductor Converter (SEPIC) and Zeta converter. The reader can examine the voltage at different parts of the circuits and perform DFT on them using the frequency response code provided in the simulation packages. The reader can find the topology and description of operation of these converters in many free online resources.

Chapter 3 described a few examples of bidirectional buck, boost and buck-boost converters. However, to demonstrate the switching strategy being used and the bidirectional nature of power flow, either the input was considered to be a dc voltage source while the output was considered to be a capacitor supplying a resistive load or vice versa. The reader is encouraged to replace the dc voltage source with a model of a battery and the output with an active load that could resemble a dc motor, to examine how bidirectional flow of power can be used to charge and discharge the battery with respect to the control of the dc motor.

Chapter 4 described a full-bridge converter, and used a few examples to illustrate the completeness of the topology as a basic functioning converter. However, the reader is encouraged to extend these simulations with details of either the sources or the system being fed. As an example, in the case of the full-bridge converter feeding an ac system represented by a resistor-inductor load, the reader could replace the resistor-inductor load with an ac grid and include the controls necessary for grid interconnection such as Phase Locked Loop (PLL). The reader can also potentially replace the dc voltage source supplying the full-bridge converter with a Photovoltaic (PV) panel and include the model to vary the output voltage according to the incident solar radiation.

Chapter 5 described a three-phase converter and the SVPWM approach to generate switching signals for the converter. However, the simulation examples merely used a resistor-inductor load as the ac system that the converter supplies. The reader is encouraged to consider designing an inductor-capacitor (LC) low pass filter at the output terminals of the converter and connect the resistor-inductor load across the capacitor of this filter. Such an exercise will demonstrate the open-loop operation of an UPS, whereby smooth sinusoidal three-phase output voltages can be produced across the capacitors of the filter.

Chapter 6 described in detail a 3-phase 3-level converter and the SVPWM algorithm that can be used to generate the gating signals. The reader is encouraged to extend the simulation to a 3-phase 4-level converter. To modify the triangle identification algorithm, the reader only needs to extend the conditions to include the outer triangles that are present in the 4-level converter. The greater challenge will be encountered in balancing the capacitors of the dc bus as there will be three dc bus capacitors. Furthermore, the inner triangles of the vector diagram will contain two or three levels of redundancy in the choice of converter voltage vectors.

7.3   The Road Ahead

As would be very clear to any reader who has reached this far, the philosophy behind writing these books differs from that of conventional textbooks. Any conventional textbook in power electronics would cover a vast range of topics. However, textbooks are written as accompanying material to university courses and therefore, they do not cover the details of a topic leaving that to the lectures and laboratories. Another category of books that can be found are the research books where researchers convert their Master's or PhD thesis into books, or quite often write books out of research projects. Though such books have a more narrow focus than textbooks, research books are written not very different to research articles that can be found in journals and conference proceedings, and specifically report the advancements made by the research team while citing those that have been used to arrive at the results. Such books are very difficult to comprehend for a newcomer to power electronics.

This book along with the previous books on power electronics have been written with the aim of providing education. However, as stated before, it is imperative that education in power electronics is revamped and made more interesting to a young engineer. This book uses basic reasoning and logical thinking to understand power electronics rather than mere mathematical analysis. Furthermore, the book heavily uses simulations and programming to explain concepts. The book uses free and open source software rather than proprietary software to ensure that any student in any part of the world can follow the book without the need to purchase software licenses. All the programs and simulations can be found on the GitHub link: https://github.com/opensourceelectrical/switching-strategies-for-power-electronics. Interested readers can also examine other related programs and simulations in the repository: https://github.com/opensourceelectrical/.

The book focuses on a very narrow topic in power electronics—namely switching strategies for power converters. The purpose of the book is to be a complete guide for any young engineer who would like to gain a working knowledge of power electronics that is relevant to industry. Furthermore, this book can also be used as training material by companies either for training new recruits and interns, or as refresher material for their practising engineers. The book has specifically been written similar to a user manual rather than an academic book in order to be of use to working people who lack the time and energy to read an academic book.

In the years to come, several more books are planned to educate young engineers on material specific to the power industry. Books are being planned in the domain of active filters, renewable energy integration, microgrids, electric vehicles and many other state-of-the-art topics. It is the author's hope that such books will popularize power electronics, while also providing industry with affordable tools to train their engineers. Interested readers are encouraged to follow the project Python Power Electronics: https://www.pythonpowerelectronics.com/.